Resuscitating America
An Independent Voter's Guide to Restoring the American Dream

C. Aaron Swisher

Copyright © 2011 by C. Aaron Swisher

All rights reserved. No part of this book may be reproduced, stored, or transmitted by any means – whether auditory, graphic, mechanical, or electronic – without written permission of both publisher and author, except in the case of brief excerpts used in critical articles and reviews. Unauthorized reproduction of any part of this work is illegal and is punishable by law.

ISBN 978-1466450134

For my nieces and nephew, in hope that we might leave you the nation and the world better than we found it.

Table of Contents

Acknowledgments i

Preface iii

America's Problems 1

Understanding Economics, Capitalism, and Why the Traditional Remedies Usually Fail 18

The Road to Ruin 33

A New Capitalism 51

A New Trade 76

Energy Independence and Environmental Sustainability 99

The Federal Budget 122

Conclusion 160

Acknowledgments

As with any literary work, there were numerous contributions of various forms that helped to bring this book to fruition. First and foremost I must thank my wife, Elizabeth, who sacrificed not only my company for many dates and social events, but provided constant encouragement and help. Dr. Benjamin Peters, Professor of Political Science at Miyazaki International College and lifelong friend, provided both editorial advice in the early stages of the book's development and encouragement and advice throughout; I could have no better sounding board for my thoughts and ideas. The presentation of my work has been greatly enhanced by my editor, Kathy McIntosh at A Well Placed Word, as well as the designer of the book's cover, Colleen Morgan. Marketing and distribution advice were provided by Jana Kemp, who was introduced to me by Vincent Martino; I am extremely grateful to both.

Preface

The American economy is a mess. While that has been painfully obvious since the advent of what pundits have termed "The Great Recession," it should have also been obvious well before that. The stock market and housing market may have done well at various times over the last decade, but employment certainly hasn't been overly abundant and real wage growth (wage growth adjusted for inflation) for most Americans has been virtually non-existent for a few decades. Despite the general feeling of wealth produced by rising home prices early in the present century, there was something eerily depressing about having to cash out the equity in your home (by accepting a higher mortgage payment) in order to pay off credit cards, take a vacation, or buy some big ticket item that you always wanted. Even now, though the recession is officially over, the official unemployment rate is stubbornly stuck around 9 percent. And if we still measured unemployment the way we used to, the rate would be more in the range of those seen during the Great Depression.

Less known is the fact that the world of professional economics is also in disarray. Few professional economists foresaw the crash of 2008, and none seem to have ideas for getting the economy going again that don't fit into the traditional liberal or conservative molds. What professional economists seem to be missing is that all the remedies proposed by the major schools of economic thought are currently in play and none of them seem to be working. Keynesian techniques of increased government spending and an expansion of the money supply are both being fully employed to no avail. Of course, most Keynesian economists would argue that we aren't doing enough, but with unprecedented levels of government expenditure *and* liquidity one would have to wonder how much more we have to strain the budget *and* the bank in order to effect a change. Economists on the conservative side make the same argument about their traditional remedy — tax cuts. It's hard to believe that the answer is more tax cuts when taxes — both corporate and individual — are already at some of the lowest levels seen in the last half-century. If any of these economic programs were working to the slightest degree one would think that the results would be more obvious.

Things aren't much better in the political world either. There was optimism that the political bickering would end with the campaigns of 2008. The election of Barack Obama seemed to usher in a new era of hopeful bipartisanship. The newly elected President seemed adamant about working with the opposition in the Republican Party. (Remember, though, that George W. Bush was elected in 2000 also claiming to be bipartisan). With three years gone since Obama's inauguration, and the two sides still split on the proper policies for resuscitating our economy, Obama's hopes of bipartisanship seem a distant memory. It's great to hope that our leaders might come together, put partisanship aside and agree on some steps that would put our country on a path of sustainable growth and fiscal discipline. But does anyone really think it's as simple as holding hands and singing "Kumbaya?" Despite the continual rhetoric about "working with the other side of the aisle," neither side ever seems to come up with a platform or a set of policy prescriptions that both sides could support.

This book is an attempt to do just that. In the pages ahead I lay out a comprehensive set of proposals that will revive the economy, balance the budget, reduce our energy problems, and lessen the damage to our environment.

Maybe most importantly, these changes lay a foundation for the solution, or at least the reduction, of most of America's social problems. I'm not talking about a foundation for a government solution to these problems, but an economic foundation that gives families the financial freedom to take the time to solve problems at home, where they should be addressed. Some of the individual proposals will seem liberal in nature; some suggestions will certainly seem conservative. I didn't plan it that way. Lo and behold, the proper set of solutions is a mix of policies that provides Democrats with their "economy that works for all" and provides Republicans with their "smaller government" and "balanced budgets."

It sounds dreamy, almost too good to be true. So before you get too excited, realize this: whether you consider yourself liberal or conservative, there will be proposals in this book that you do not like. There are policies in this book that I, the author, do not like. I didn't choose this mix of proposals to make people feel warm and fuzzy; I didn't choose them to create a best seller; and I certainly didn't choose this mix to appeal to an electoral "base" that I need to win over in order to be nominated for some high political office. These policies were chosen because they make sense. If you're not willing to open your mind to an idea you don't like, don't read any farther.

That brings up another vital point: this is a comprehensive package. The policies suggested in the pages ahead aren't an "ala carte" menu that we can pick and choose from to suit our particular tastes. Enacting some, but not all of the policies could possibly do more damage than good. None of America's problems exists in a vacuum. The problems that our nation faces are a spider web of interconnected issues and must be addressed with a set of similarly connected policies, developed with regard for one another. Our energy policy is affected by and has an effect on our defense policy and our fiscal health. Taxes are affected by defense spending and have an effect on social norms and culture. Income distribution problems are affected by international trade and, in turn, won't be easily fixed without considering tax policy as well as the money supply. Far too often our leaders present ideas that are great for solving one specific problem, but create havoc in other areas of our society. The goal of this book is to put forth a set of solutions that allows for the resolution of all these problems, and to do so in a way that will find broad, bipartisan (or hopefully non-partisan) support throughout the country.

In order for that to happen, our country must find a solution to its biggest economic problem. That problem, and how it relates to many of our other problems, is laid out in Chapter 1. The factors that have brought about this predicament are explained in Chapter 2. Unlike most of today's economic works, I have avoided a long-winded review of the specific details that have marked the economic landscape of the last few decades and chosen instead to review the fundamental aspects that underlie our capitalist, free market economy. Chapter 3 gives an analysis of the typical Republican and Democratic responses, and why those responses usually fail to fix the economy and provide long-term prosperity.

Chapters 4 and 5 present a comprehensive set of measures that we can take to revive the economy and lay a foundation for real future economic growth. That economic foundation will be of limited use, however, if America fails to develop a long-term energy strategy and neglects to get its fiscal house in order. Therefore, Chapter 6 resolves our country's endless thirst for energy and resources in a way that complements the economic and fiscal policies offered in earlier chapters. The typical solutions put forth by our dominant political parties — laden with tax credits and business subsidies — are rejected, and replaced with simple, market-based solutions that make more sense. Chapter 7 suggests meaningful ways to cut government spending and bring about a balanced budget. Because these suggestions transcend the typical political rhetoric, they make the choices necessary to reduce the size of government while not leaving vulnerable Americans at risk.

I will note again that not everything you read in the following pages will be to your liking. Any athlete training for serious competition dislikes part of his/her training regime, but each part is critical. And if our country is to take advantage of its full potential, we too will have to endure some distasteful exercises. But if I have done my job properly, by the end of this work you will see how all these policies could work together to truly rescue this nation from the edge of the economic and fiscal abyss where it presently seems perched. There are common sense solutions to the seemingly intractable problems that our dominant political parties seem unable to solve, and those solutions need a voice. My hope is that this book is that voice.

America's Problems

If someone asked you what the number one problem in America is today, what would you say? Crime? Abortion? Illegal immigration? The economy? Maybe corporate scandals or political polarization? What about the exporting of American jobs? Threat of attack from North Korea or Iran? If you regularly attend Tea Party rallies, you'd probably say taxes or government spending. But America's *biggest* problem is really none of these.

What if I told you that the greatest threat to our country and our way of life is income inequality? I'm not talking about income inequality in the sense that some people simply make more than others, but income inequality in the sense that a very small percentage of Americans — really less than 1 percent — make outrageously large incomes, while the vast majority of Americans end up underpaid and struggling to make enough to scratch out a living.

Consider this: in 2007, the top 1 percent of households took home 23.5 percent of the country's income, and the top 10 percent took home 49.7 percent (up from less than 35 percent in 1980). That means 90 percent of the country's households were fighting over roughly 50 percent of the country's income. And as the evidence shows, the real concentration of income is located *within* the top one percent. In 2007, the top .01 percent of households (one out of every 10,000) took home over 6 percent of the country's income.[1]

Killing the Economic Engine

It should be obvious to anyone who understands real world economics that widening income disparity is a problem for the economy. If it's not, let me use an analogy to demonstrate the point. Think about the internal combustion engine in your car. The engine produces power through the work of pistons that move up and down inside cylinders, creating the force needed to propel the wheels. To reduce friction within the cylinders, the pistons are lubricated with oil. This oil sits in a pan in the bottom of the engine and is equally available to all the pistons. If the amount of oil in the pan is insufficient to lubricate the pistons, friction develops and the engine eventually breaks down (or seizes up and stops). In order to prevent that from happening, more oil must be added to keep the engine running smoothly.

Now imagine that, for some reason, a majority of the oil in your engine got stuck in one piston (and thus isn't available to the other pistons). The piston with the abundance of oil would be well lubricated and operating sufficiently, but the other pistons would begin to experience friction and start to break down. If the pistons lacking oil begin to fail, the entire engine will shut down (one piston simply can't do it all). More oil could be added to the engine to lubricate the other pistons, but this would mean that the overall engine would have too much oil in it, causing other problems.

Our economy operates in the same manner. The lubricant of our economic system is money. The pistons, which produce the power of our economic engine, represent workers and consumers in various segments of society. The entire engine is our economy. From time to time, our economy suffers from the same problem described above regarding the internal combustion engine: too much of the lubricating agent — money — becomes stuck in one piston (or in a small segment of our society) and is therefore not available to the other pistons.* Although the other pistons may be working, their existence becomes difficult because their lives don't have the appropriate amount of lubrication (i.e., they don't earn enough to function properly in the economy). This leads to a breakdown of the overall engine — what is commonly referred to as a recession, or in severe cases, a depression.

If we look back over America's economic history for the last century or so, the evidence is fairly obvious that this is the problem with our economy. After all, America hasn't always had an income disparity problem. You see, our country has gone through both periods of great inequality as well as periods with a more equitable wage distribution. What has developed over the last few decades is really the re-emergence of a problem the United States has been through before. And the difference in experiences between those various periods — for our economy and our society — makes the case for a better level of equity.

Prior to World War II, income disparity in America was a function of wealth ownership. Tax return data from the first few years of the income tax show that income earners at the very top of the wage scale tended to categorize themselves as "Capitalists: Investors and Speculators."[2] Since the vast majority of workers were paid very little, businesses produced large profits. These profits accrued — mainly in the form of dividends — to the small segment of society that owned those business assets. People like Andrew Carnegie, J.P. Morgan and John D. Rockefeller enjoyed enormous incomes, not because of bloated work salaries, but because they owned large business conglomerates and those enterprises enjoyed abnormally large returns. Regardless of the source, the effect that the vast level of income disparity had on the economy was obvious.

Because incomes were spread so unevenly — with a small few taking enormous rewards and the overwhelming majority of workers struggling to get by — the economy continually sputtered in the latter half of the 19th century and the first part of the 20th. At that

---

* Economists tend to believe that money never pools in one segment of society, because that segment makes it available to other segments through investment and credit markets. We'll see why this isn't always the case in the next chapter.

time, most Americans worked in the agricultural sector, where wages were low and debt was common; many workers lived payday to payday. Any time financial disruptions disturbed the delicate fiscal balance of the agricultural industry, the country would experience a massive depression or panic, such as those in 1873 and 1893. Even as the country began to industrialize, the low wages that most workers made kept the economy fragile. Between 1910 and 1927 alone, half a dozen recessions wracked the United States.

Then, in 1929, after a growing speculative bubble on Wall Street, the stock market collapsed and the economy soon went into a tailspin. At the time (1928), the top 1 percent of households took home about 24 percent of the country's income and the top decile nabbed around 49 percent — numbers eerily identical to what we saw prior to this latest economic meltdown.[3]

The reaction of the Roosevelt Administration to the depression was a mix of programs and regulations aimed at getting the country back to work and helping it avoid another financial catastrophe. While the initial programs helped, they really weren't enough to overcome the economic problems that had been caused by such an inequitable distribution of the country's income. The programs did little, if anything, to correct the income imbalance in the economy. And so, the United States spent most of the 1930s simply trying to regain its economic footing.

This changed in the late '30s and early '40s, however, for a number of reasons. First, the establishment of a minimum wage (1938) put a floor under the lower end of the wage scale. Second, our entry into the Second World War lifted wages by increasing labor demand (jobs created through war spending) and reducing labor supply (sending men off to war). The Great Depression and World War II also caused a reduction in the wealth holdings of the elite (remember, this is where high-income individuals were getting most of their earnings). Finally, a progressive tax system and an increase in unionized labor added to the great compression* of wages.

---

* "The Great Compression" was actually a title used by economic historians Claudia Goldin and Robert Margo in an August 1991 paper to describe the lessening of income inequality during this part of the 20th century.

This more equitable wage structure facilitated a vibrant economy where more Americans prospered: an economy less prone to recessions, panics, and the other problems that America had experienced up until that time. The advances were so widespread that *Time* magazine reported in 1953: "Even in the smallest towns and most isolated areas, the U.S. is wearing a very prosperous, middle-class suit of clothes.... People are not growing wealthy, but more of them than ever before are getting along."[4]

To get a broad sense of how the economy acted over this time period, look at Table 1 below. It shows periods of economic recession (in black) in the 100 years from 1870 to 1970. Notice the frequency of recessions prior to 1940 and compare that with the segment from 1940 to 1970. Not only are the recessions more numerous, they tend to be of greater length than those experienced later in the 20th century. While there are some economic downturns during this latter period, they tend to be less frequent and very short in duration. In other words, the economic engine was running better because oil was making it to all of the pistons.

Table 1

Recessionary Periods from 1870 to 1970

Unfortunately, in the '70s three things happened. The first was that income disparity began to increase. Wages for those in the upper income classes started growing at a rate that kept them well ahead of inflation for the next three decades. The average salary, adjusted for inflation (2006 dollars), for someone in the top 1 percent of income earners grew from $239,418 in 1974 to $653,365 in 2006 — a total growth rate of 173 percent, or almost 3.2 percent a year. Average salaries for those in the top .1 percent exploded, going from $531,759 to $2,617,596 over the same time period — growth of over 5.1 percent per year.[5]

At the same time, income for the typical American stagnated. The inflation-adjusted minimum wage (2007 dollars) fell from a value of $8.42 per hour in 1974, to roughly $7 per hour in 2009. Average hourly earnings for production and non-supervisory workers at private, non-farm businesses didn't fair quite as badly, but still fell from $8.99 an hour (2007 dollars) to $8.68 over the same time period.[6] The median income did better, but not by much. In 1974, the annual median income was $42,459 (2007 dollars); by 2008, it had risen to only $48,443.[7] And even that's not such a positive development when you consider the fact that Americans were putting in more hours to earn that income in 2008.

Today, income disparity is in large part due to the oversized incomes of an elite few and the undersized incomes of the vast majority. To see what I mean, consider the pay ratio of an average chief executive officer to that of an average worker. In 1965 the ratio was 24 to 1 (the average CEO made 24 times that of an average worker); in 2007, the ratio had grown to 275 to 1.[8] And even CEO pay looks paltry compared to that of private equity and hedge fund managers. In 2006, the average pay of one of these money managers was $657.5 million, more than 16,000 times what an average worker made that year, and roughly 61 times greater than the average CEO.[9]

Given America's experience with income disparity near the beginning of the century, one might expect that this period from the early '70s until today would be filled with more frequent, and longer, economic recessions. But that actually isn't the case. The reason is due, in large part, to the other two changes that occurred in the '70s.

In 1973, Richard Nixon took the United States off the gold standard. With the value of the U.S. dollar no longer tied to gold, the amount of money in the economy could easily expand, creating a cushion against economic downturns. Since that time, the level of money in circulation in the economy has increased dramatically. With the exception of a brief period in the early '80s, every time the economy has started to sputter — because too much money was starting to collect in the hands of a relatively small number of citizens — the Fed[#] has added more "liquidity" (i.e., money) to the economy. Look at Table 2 below. Not only does it show periods of economic recession (in gray), it also shows the Adjusted Monetary Base of the United States. Notice how the monetary base is relatively flat until the early '70s, when it begins an ever-increasing climb to the present day. An increasing money supply keeps interest rates abnormally low and encourages borrowing, thus giving people who might not be earning enough money through their paychecks a ready supply of cash. This continual increase in funds was necessary to make up for the worsening income inequity we've seen since the early '70s. In other words, the United States has managed to avoid recession-prone stretches during this period by continually increasing the money supply (i.e., dumping more oil into the sputtering engine to keep it running).

Table 2

Recessionary Periods from 1920 to 2010 with Adjusted Monetary Base

The other factor that helped to keep the economy afloat over this period was the continually increasing budget deficit. The United States went from a budget surplus in 1969 to budget deficits of almost 6 percent of gross domestic product (GDP) in 1983. In fact, in the 20 years from 1975 to 1995, the budget deficit dropped to less

---

[#] The Federal Reserve (or, "the Fed") is the entity in charge of controlling the money supply and interest rates.

than 2 percent of GDP only once. Compare that to the 27 years from 1947 to 1974, when the deficit topped 2 percent of GDP only once. This additional spending, beyond what the government was collecting in taxes, has provided a strong stimulus that has helped keep the economy afloat.

So could we say that the solution to our problem is really to keep expanding the money supply and increasing the national debt? No, not at all! To begin with, neither of these are sustainable ways to keep the economy moving. Each of these "solutions" causes economic problems of its own. More importantly, neither of them really fixes the true problem; they simply mask the symptoms. Hence, even the stimulative effects of additional money and budget deficits haven't stopped the inevitable: our economy is presently suffering through an economic recession large enough to invoke comparisons with the Great Depression of the 1930s.

But before the economic engine even gets to the point of a total breakdown, large levels of income disparity cause other problems in our economy.

When a majority of Americans have incomes that fail to keep up with inflation, their purchasing power decreases and their spending habits change. This is a huge problem, particularly for a country that tends to look to "the market"—individuals making rational (or semi-rational) choices, driving change and creating solutions—to solve its problems. How can we expect the market to work properly if most citizens are too poor to make the choices they'd really like to make? How can the majority of Americans drive the market in a given direction if their choices are continually being distorted by a lack of income? "Buy American", and thus keep jobs here in the U.S.? Or save a much-needed dollar by buying a Chinese import? Spend a little more for healthier, or environmentally friendly, products? Or live within your paltry budget and hope the money-saving decision doesn't decimate your health, or ruin the environment you live in? Many Americans are simply limited in the number and kinds of choices they can make because of their relatively limited incomes.

Low wages obviously force consumers to be extremely cost conscious, looking for the lowest price in every transaction. And if a majority of consumers are in this position, that is what the market will cater to, reducing price by any means necessary. This creates an atmosphere where companies start to churn out inferior products at the lowest possible cost, rather than trying to strike the appropriate balance between price and quality (i.e., value). Finding a high-quality, long-lasting product thus becomes more and more difficult.

This cost-conscious trend also forces businesses to hold the line on labor costs, denying raises and skimping on cost-of-living increases. If companies can't keep wages low enough in the United States, they may outsource work to other countries where labor may be cheaper and production can be completed at a lower cost. This process of exporting production jobs and importing manufactured products expands our trade deficit, weakens the U.S. dollar, and increases unemployment in the United States. It also reinforces the income disparity problem by strengthening the cycle of falling wages for typical Americans.

As real income for the average American falls, saving for things like retirement, or a child's college fund, becomes more difficult. Large purchases, such as an automobile, must be made by borrowing money. Even smaller purchases—washing machines, dryers, furniture—must be made on credit. These developments have transformed us into a nation of debtors. Properly managing one's money has always taken effort, but as wages have stagnated or fallen for most Americans, it has become almost impossible.

All of these problems have an additional effect on government budgets. The loss of manufacturing jobs has a negative effect on state and federal budgets by eroding the tax base and increasing the number of people receiving unemployment benefits. Even for those who manage to keep their jobs, the falling standards of living they experience cause many of them to seek assistance, usually from the government. Student loans for college, tax credits for having children, food stamps, the Earned Income Tax Credit and various entitlement programs (to name just a few) have all been created, or expanded, to help make up in areas where the typical American wage falls short. In 2010, 35 percent of wages and salaries consisted of government payouts — Social Security, Medicare, unemployment, and other forms of social welfare. If you think that most of that was due to the Great Recession, consider the fact that even the pre-recession level was 26 percent. As a comparison, in 1960, when income was distributed more equitably, these payments represented only 10 percent of wages and salaries. And whether the assistance comes from local, state, or federal programs, having greater numbers of people on public assistance plans puts a strain on government finances and increases the chance that these entities will either run a budget deficit, have to cut other important programs, or have to increase taxes to make up the shortfall.

Simply consider the national debt. An enormous portion of the nation's future debt is projected to be social programs like Medicare, Medicaid and Social Security. These programs were started to provide a safety net for Americans — to ensure that all citizens of this country had minimum standards of health care and old age income, regardless of the unforeseen difficulties they encountered. But increasing levels of income inequality mean that more working Americans are relying on these programs to provide their health care and retirement income. The more these programs are needed and utilized, the more they cost and the harder they are to reform.

Sometimes income disparity causes a multi-pronged problem, as is the case with Medicaid. Poorer Americans tend to have diets that are less healthy than wealthier Americans. This means that poorer Americans who would probably qualify for programs like Medicaid also tend to need more health services throughout their lives than their wealthier counterparts. So for entitlement programs like Medicaid, income disparity is actually a double whammy — more people qualify, and those who do use it more than they would if their income provided a better standard of living.

The situation is similar for programs like Social Security. Since higher income Americans only pay Social Security taxes on the initial portion of their earnings (the first $106,800 in 2009), increases to their income don't translate into additional revenue for the Social Security system. On the other hand, if more of the country's income pie went to poor and middle class wage earners— whose payroll taxes increase as their income increases — an increase in income would mean more revenue and stability for this government program.[#] Higher wages on the lower end of the pay scale would also allow many of these workers to save for their own retirement and reduce their dependence on the Social Security

---

[#] This is not to suggest that a lower level of income disparity would completely fix the Social Security program (a subject I'll touch later), just recognition of the fact that income disparity causes problems that are less than obvious to most people.

program. This would make it easier to either eliminate the program, or enact changes to guarantee its long-term solvency.

All this debt, of individuals as well as the government, is a byproduct of the majority of Americans trying to maintain a constant standard of living on falling levels of income. And as the debts grow larger and larger, the day of reckoning looks worse and worse. This is because any time entities (whether individuals or governments) borrow money to fund current consumption, they rob from future consumption. Not only must the debt be paid back, it must be paid back with interest, so future consumption has to be reduced from what it could be in order to pay off the over-consumption of today. Consider the fact that American consumers have over $2.4 trillion worth of debt (not including mortgages)[10] and our government is in hock for another $14 trillion. That's a lot of consumption (with interest) that must be sacrificed by our economy in coming years, or even decades. And much of this can be blamed on the fact that fewer than 1 percent of our countrymen have been taking a larger and larger share of the income pie, leaving less and less for everyone else.

If it's not already blatantly clear, large levels of income disparity do not bode well for our economy or our government's finances. Economist Ravi Batra might have put it best in his 1996 book *The Great American Deception*:

> History shows that extreme inequality is the bane of all economic systems and civilizations. In ancient Egypt, Mesopotamia, Rome, India, China, Europe, and Japan, gross disparities of income and wealth demolished what were once rich and cultured societies. At best, enormous inequality spawns peaceful rebellions; at worst, bloodshed and violence.
>
> In medieval times, feudalism vanished because of inequality. Later, so did Tsarist Russia and Imperialist China. ... In modern eras, wealth disparity has engendered speculative bubbles, poverty, and depressions.[11]

It's interesting to note that since that publishing, America has seen two of the bigger speculative bubbles in its history (tech and housing)*, is suffering through its worst economic downturn since the Great Depression, and is starting to show signs of political instability. *Large levels of income disparity just don't produce a properly functioning society!*

A Problem for All Problems

But as the quote above implies, extreme income inequality is a problem that stretches far beyond a country's economic and financial stability. It influences our lives and our society in numerous, unobvious ways — affecting everything from the morals and ethics of our country's citizens, to the very foundations of our government. Our income disparity issue is at the center of a web of social problems that affect all aspects of our society.

Let's start with the family unit. Money problems are at the heart of marital problems. A 2010 report by the National Marriage Project found that "Middle American" couples (the 58 percent of adults who have a high school diploma, but no four-year college degree) are increasingly foregoing marriage, experiencing increasing divorce rates, and having more children out of wedlock. The group states that "data indicate that trends in non-marital childbearing, divorce and marital quality in Middle America increasingly resemble those of the poor, many of whose marriages are fragile."[12] Even for couples that marry and manage to stay together, the result is often that both spouses need to work in order to make life viable for the family. The additional hours spent by the second spouse working to provide needed income rob the household of much-needed labor time at home. With both spouses working (or with one parent trying to make it as a working, single parent) children are usually put in day care and daily chores are usually handled in the evening, during the time once spent as a family. The result is less time spent with the kids and a potential weakening of family bonds.

As more and more family units break down, societal problems increase. The National Marriage Project points out that the breakdown of marriages "increases the odds that children from Middle America will drop out of high school, end up in trouble with the law, become pregnant as teenagers, or otherwise lose their way."[13] Increased crime rates, gang membership, drug use and teen

---

* There is some speculation that we're entering our third bubble – commodities.

pregnancy all tend to be the products of poverty, broken homes and children who fail to get the proper love and attention they need from busy parents. These are problems that not only affect an individual household, but our entire society.

Lower wages also have an effect on decisions such as whether or not to abort an unborn child. The abortion rate for women living below the poverty level (roughly $10,800/yr. for a single woman with no children) is more than four times that of women who earn at least 3 times the poverty level (roughly $32,400/year or more).[14] And three-fourths of women having an abortion cite the inability to afford a child as one of the reasons for the procedure.[15]

The need for both parents to work in a typical household has also contributed to the break down of a once healthy American diet. In the mid-1940s, when a typical household needed only one breadwinner, 40 percent of the vegetable produce in the American diet came from a family garden.[16] This not only saved the family money (by growing their own food), but it produced a much better diet than the average American has today. The appropriate quantities of phyto-nutrients, vitamins, and minerals — gained by eating food picked from the backyard garden — are now sadly lacking in the average American diet. Of course, tending a family garden takes time, time that a household with two breadwinners doesn't have. Since time is scarce for parents, processed, pre-packaged food has become the norm for breakfast, lunch and dinner. These food sources have more fat, more preservatives, more sugar (or sugar substitute) and more sodium than balanced meals made by hand with whole, fresh foods. The results are wide ranging, from obesity, infections and cancers, to depression and reduced self-esteem.[17]

If this weren't bad enough, the financial stress that comes from barely making ends meet has been linked to depression, hair loss, diabetes, heart disease, sexual dysfunction, ulcers, cancer, and more. These health issues are not only problems in their own right; they increase absenteeism at work and drive up the demand for healthcare. This, in turn, affects business profitability, increases healthcare costs and reduces one's quality of life.

But these aren't problems that have to be dealt with for America's elite income earners. Their income has stayed so far ahead of inflation that they can easily maintain, or even increase, their standard of living each year, without going into debt. Their earnings leave them with plenty of excess funds to do with as they please, even after their needs and wants have been met. But as we'll soon see, this too causes a raft of problems — for individuals as well as for society.

As the living standards of the rich and famous rise, our society's idea of what it means to "live the good life" also rises. The definition of "success" becomes outdated and replaced with a new image that is kept up-to-date by our constant view into the lives of America's upper crust through TV and magazines. The continual climb in living standard of the upper echelon of society gives us an over-inflated sense of how our lives should be lived. We've seen the way the super-affluent live, and we desire that life for ourselves. But if the lives of the upper crust weren't so extravagant, maybe the lives of the common class wouldn't seem so ordinary. If the income distribution of our country didn't give a limited few more money than they need and the vast majority less, maybe we could all settle for lives that are a little more satisfied and sustainable. Maintaining a constant living standard with a stagnant or falling real income is difficult enough; trying to keep up with the ever-improving lives of the wealthy has most American families frustrated, working longer and harder, and still swimming in massive amounts of debt.

All this anxiety and unhappiness eventually spills over into the political realm. The degree of political polarization that America is now experiencing can be directly attributed to our level of income inequity. Economist Paul Krugman, citing a detailed study by Nolan McCarty, Keith Poole, and Howard Rosenthal, has pointed out that as income inequality increases, political polarization becomes worse and producing a political solution for any other problem becomes more difficult.[18]

Our income inequity is also corrupting the very foundations of our democracy, allowing the greed of a few to destroy the opportunities of the many. Because our present political system is extremely reliant on, and influenced by, campaign contributions, money has become power (if this wasn't the case before). This means that the small minority of Americans who have outrageously large incomes wield much more power in Washington than the massive majority of Americans whose incomes don't allow them to play the game.

The outcome is less than ideal. Simply look at any piece of legislation passed in the last three decades that has caused major economic trouble for our country. What you find is a bill (or an addition to a bill) pushed by moneyed interests whose goal is to create a system or an environment that allows them to make even greater returns. The Gramm-Leach-Bliley Act (1999) repealed the Depression-era Glass-Steagall law that prevented banks from getting involved in investment banking and insurance operations. That contributed to the massive bank losses and subsequent bank bailouts we saw in 2008. Those bank losses, in large part, were also caused by unregulated derivative trading permitted by the Commodity Futures Modernization Act (2000). Both of these bills gave the banking industry the ability to earn massive profits. Unfortunately, they were both destabilizing for our economy, and when things went sour, the taxpayers were left with the cleanup fee. Legislation like this isn't pushed by ordinary citizens (sometimes it passes without most Americans even knowing it); rather, it is often a result of lobbying by large corporations and extremely wealthy individuals.

In addition, our elected officials tend to come from the wealthier ranks of citizens as well. That's a problem. People who are wealthy typically view the world much differently from those who are less well off. That doesn't necessarily mean they're bad people, but it does mean they probably view things from a different perspective than the majority of Americans. If decision-makers (in government *and* business) are experiencing and viewing the world from a much different set of circumstances, how good will their decisions be for the multitude that have to live with their decrees? Shouldn't our leaders (or at least *some* of them) come from the common ranks? Or at least have to live in circumstances that are at least *somewhat* similar to our own. It's not hard to look at the legislation that comes out of Washington and notice that quite a bit of it is focused more on helping large corporations and the wealthy than on charting a wise, long-term course for our country and economy.

Conclusion

Imagine what life would be like in America if every worker earned the same wage. In 2008, the average household income in the United States was $68,424.[19] Think about what life is like for families with an income in that range—not enough to be instantly rich, but certainly enough to live better than most Americans are currently living. Now imagine a family enjoying that living standard from the earnings of just one bread winner. If you really take the time to envision it, you'll suddenly understand the damage our growing income disparity level has caused.

But we can't all have the same income, nor should we. Incentives provided by disproportionate income levels are essential for the proper functioning of our economic system. The potential for greater reward spurs the creativity, inventiveness, and stronger work effort that improve the overall living standard of our society. And there will probably always be a segment of our society that is poor, whether due to lack of effort or poor financial management skills. This too provides needed incentives for extra effort and improvement. In other words, a certain degree of income disparity is absolutely essential.

Unfortunately, income disparities as gaping as those we presently have create big problems. When common, hard-working people can't make a living through legitimate means, society's list of ills is sure to multiply. And while I've listed many of those troubles here, this chapter's litany was far from exhaustive. The point is this: countless problems in America have been caused, or worsened, by our vast and growing inequality in incomes.

That's why we need a comprehensive solution, one that not only solves this fundamental problem, but also provides a foundation from which all our other troubles can be addressed. It makes no sense to try to come up with a resolution to any of the nation's issues until the underlying cause of a disproportionate number of these issues is resolved.

I'd be a fool to try to tell you that solving the pay equity problem in America will suddenly solve all of our ills and create a grand Utopia. It simply won't happen like that. Our culture and social norms have changed in some negative ways that will take generations to rectify. Getting our society back to a point where cohesive family lives and producing savings rather than debt are commonplace will not happen overnight. Income inequality isn't America's only problem, nor is rectifying it the only solution.

But what we can do — and should do — is foster an environment where these changes are possible. We can do this by establishing a system that encourages a more equitable distribution of the country's income. At the same time, we can provide leadership and incentives that direct us back to a more moral and sustainable way of living. If done properly, solving the income disparity problem could reduce the magnitude of nearly every other problem in America and foster an environment where people from all political spectrums can come together and work toward the best solution for each issue. Finding a fix to the wage inequity in this country won't solve all our problems, but few of our problems will truly be solved without one.

Understanding Economics, Capitalism, and Why the Traditional Remedies Usually Fail

It's not enough to know that our biggest problem is the massive disparity between incomes of the ultra-rich and those of the majority of Americans. If we are to solve this problem, and do so in a way that is logical and fair, we have to understand why this phenomenon develops and why it creates problems for the economy. In order to do that, we have to come to a basic understanding of the free market capitalist system on which our economy is based. Doing so will not only help us understand why extreme levels of income disparity develop, it will provide a foundation for analyzing the responses of Democrats and Republicans to this trend and the economic havoc it creates (in the next chapter). It will also offer a basis for developing economic ground rules that can lead our country back to prosperity.

How a Free Market Capitalist Economy Works

The basic tenet of capitalism is that the use of capital (tools, machinery, etc.) makes workers more productive and thus produces greater profits. For instance, suppose a company was hired to dig a large ditch. The company could hire 100 workers to dig the ditch with their bare hands. Or, the company could employ 20 workers with shovels (capital equipment) to dig the ditch. Better still, if the owner of the company owned a backhoe (more advanced capital equipment), he might only need employ two workers who could probably get the job done faster than 100 men with their bare hands or 20 with shovels. Thus, as more advanced capital is used, the amount of revenue from the job that is turned into profit increases (versus what is paid as wages for laborers).

Capitalists believe that this extra profit gained from the use of capital equipment should go to the owner of the capital. Having the additional profit accrue to the capital owner provides an incentive for employing capital. As more tools and machinery are employed, workers become more productive and the society's standard of living increases. Thus capitalism produces a more affluent society than other economic systems.

Now let's add to this capitalist philosophy the free market economic theory. Proponents of the free market model believe that the market is the best determinant of prices — versus the government, or some other entity — because transactions in a market result in fair and efficient prices. Of course this expected product of "fair and efficient prices" is based on the concept of a purely competitive market. This concept, however, requires us to simplify the complex economy we live in by making a number of assumptions. Some of these basic assumptions are:

1. There are many buyers and sellers (or consumers and producers) in the market, and so no one buyer or seller can exert power or control over the market.
2. There is always full employment and workers move effortlessly from one occupation to another.
3. All market participants (both consumers and producers) have perfect knowledge.

We'll focus on the first two assumptions for now and come back to address the third one in a later chapter.

Many Buyers and Sellers

Under the theoretical free market model, there are many buyers and sellers in every industry or market. Therefore, no single buyer or seller has power over prices in the market. If a business or business owner begins making profits that are excessively large, more entrepreneurs will move into that line of business (more competition), driving down profits to an optimum level.[#] In other words, an abundance of competition holds down prices, as well as profits.

What we find in the real world, particularly in America, is a situation vastly different than the free market model. There are usually a limited number of producers in the marketplace, but many consumers. Consider, for example, how many cereal producers there are in the United States. If you consider the number large, compare it to the number of people who eat. Or consider the meat packing industry, where four companies control 84 percent of the beef market,[20] driving prices up for consumers and down for ranchers who supply the beef. Any objective analysis of American industries would reveal that real, robust competition is usually lacking.

When supply is limited — as it could be with a small number of sellers — prices rise. So when the number of producers is limited, it gives those sellers (producers) a pricing advantage. An extreme case of this would be a monopoly, when a single producer has ultimate pricing power over consumers. However, even a small amount of pricing power gives producers the ability to earn what economists call "excess profits," or profits that are above the optimal rate of return we would see under purely competitive conditions.

Permanent Full Employment

The free market model also assumes that there is no unemployment and laborers move effortlessly from one occupation to another. Since it is extremely difficult to measure the exact effect

---

[#] "Optimum" here is from the viewpoint of the society, not the business owner. Obviously, a business owner who has had his profit level reduced would not refer to that as optimal.

of unemployment on wages, the creators of the theory just assumed there was no unemployment. This lack of unemployment ultimately means that wages are higher in the theoretical model than they are in the real world. Because workers in the real world are forced to compete with an overabundant labor supply, there is essentially no floor under their wages and their pay may fall with little or no restraint.

Realize that in the labor market the "producers/sellers" are actually workers, producing labor, and the "consumers" are corporations and businesses, consuming labor. Having numerous producers of labor and a limited number of buyers of that labor, gives the pricing advantage to those buying the labor (i.e., businesses). Having any degree of unemployment — excess labor — gives a disproportionate and pronounced advantage to employers in the market to buy that labor. Note that this is also why labor unions are so effective at raising the wages of working people. Any time you can limit the number of people supplying labor, or get those numerous laborers to act as one supplier, you transfer pricing power away from companies and back to laborers.

How the Ultra-Rich Get to Where They Are

Just analyzing these first two assumptions shows us how different the theoretical market is from reality. Since businesses have more pricing power over the goods and services that they are selling *and* more pricing power over the wages that they pay their employees, they are able to earn a level of excess profits that would be unthinkable in the abstract world of the "perfect" free market. This means that the country's money tends to collect in the hands of businesses and business owners, while wages for most American workers tend to be lower than they should be.

This wouldn't be a problem if all, or maybe even most, Americans had an ownership stake in our country's businesses. The abnormally low wages that are earned by workers in general would end up being supplemented by the overly large profits of the business community, both of which stem from the same imperfections. This has never really been the case; America's wealth and asset ownership has always tended to be even more concentrated than our income distribution.

As mentioned in the first chapter, this accrual of excess profits, paid out as dividends into the hands of the relatively few individuals who owned America's business assets, caused the vast amount of income disparity that the country experienced in the late 19th and early 20th centuries. And while businesses don't pay out dividends to the degree that they once did, our recent level of income disparity is similar to that witnessed just before the Great Depression. This means that the cause or our income disparity is slightly different than in previous generations. Allow me to explain.

It's obvious that as a person works more hours he or she gains more income. Those who are better educated, more creative, more talented or more productive tend to earn higher incomes as well. Additionally, as one takes on more responsibility in the workplace, he or she tends to be compensated even more. In other words, there tends to be a direct, although not always consistent, causal relationship between work and reward.[+] But if we follow the income development of those who make extremely large incomes, what we find is not such a steady, incremental progression. What we see is a gradual rise in income — by virtue of education, effort, responsibility, productivity, etc. — *up to a point*. Beyond that point, reward takes off like a rocket, increasing at a rate that effort and education never could. This is because the person's income is now based more on other forces, and most of those are out of his or her control.

There are numerous reasons for this. Robert Frank and Philip Cook laid out the most plausible one in the mid-90's in their book *The Winner-Take-All Society*.[21] Frank and Cook convincingly argue that the runaway growth in the income of such a small segment of Americans is due to markets in which the most effective producer reaps an overly large portion of the rewards. In other words, even though a minimal difference may separate the first and second place performers in a given competition, the winner walks away with almost all of the rewards. Consider two Olympic sprinters, for example, who have both trained as hard as possible. During the medal round, one sprinter beats the other by .01 of a second, taking the gold. Although there is hardly any difference in their performances, the gold medal winner will end up with millions of

---

[+] It doesn't take one long, working in the corporate world, to figure out that a person's compensation doesn't always correlate to talent, responsibility, or even work effort, even for middle-class income earners. For the purposes of this discussion, though, we'll assume that this relationship holds true, at least to a point.

dollars in endorsements; the silver medalist will hardly be remembered. Or consider two actresses trying out for their first movie role. Although they both may be equally talented, only one can be chosen. The one chosen for that initial film may go on to enjoy blockbuster status and earn millions of dollars a year, while the unselected actress barely scrapes by performing masterfully in the local theater.

Traditional economics argues that each wage earner in a society is paid according to what he or she produces. But such large disparities in income between two producers that have such a minimal difference in performance show that the traditional economic view fails in such "winner-take-all" markets. Or consider the instance when an author, such as Stephen King, is paid a multi-million dollar advance for a book that hasn't even been written. If pay is for production, how does traditional economic theory explain payments that are made before anything is produced? And these winner-take-all characteristics appear not only in athletic, entertainment, and literature labor markets, but in the market for high-profile lawyers, doctors, sales people, and investment bankers as well (to name a few).

In addition to winner-take-all market effects, the success of many of America's top wage earners can usually be attributed to other circumstances that are beyond their control. Author Malcolm Gladwell demonstrates in the book *Outliers* that "extraordinary achievement is less about talent than it is about opportunity." In chapter after chapter, he shows that those who have achieved greatness, or would be expected to achieve greatness, did so with societal help or with the help of a great dose of luck. While Gladwell admits that great achievers have contributed great effort, the rewards they receive are also due in large part to factors well beyond their control.

Outliers provides evidence that one's success in sports like hockey and baseball depends in large measure on the month of the year in which one was born, because of the way young athletes are chosen for developmental league teams. It explains how much of the foundation for the success of billionaires like Bill Gates, John D. Rockefeller, Andrew Carnegie, and others, came from events and circumstances well beyond their control. A perfect example is the admission of students to Harvard University. Gladwell writes:

"...In 2008, 27,462 of the most highly qualified high school seniors in the world applied to Harvard University. Of these students, 2,500 of them scored a perfect 800 on the SAT critical reading test and 3,300 had a perfect score on the SAT math exam. More than 3,300 were ranked first in their high school class. How many did Harvard accept? About 1,600, which is to say they rejected 93 out of every 100 applicants. Is it really possible to say that one student is Harvard material and another isn't, when both have identical – and perfect – academic records? Of course not."[22]

Now imagine two students intent on becoming lawyers, both with perfect verbal and math scores on their SATs. One may be accepted to Harvard and one may not. And yet a degree from Harvard may mean the difference between earning a million dollar salary working for a Wall Street financial firm and a $150,000 salary at a run-of-the-mill attorney's office. Can we really say that there is an $850,000 a year discrepancy between two people who both had perfect SAT scores, simply because one was fortunate enough to be accepted to Harvard University?

But "winner-take-all" market effects and "outlier" factors aren't the only reasons some incomes have gotten ridiculously out of control. Sometimes plain old greed and manipulation have played a role. This is most evident in the salaries of CEOs of publicly held companies.

In 1993, in an attempt to slow the rapidly growing salaries of corporate executives, a law was passed that only allowed corporate deductions of CEO pay over $1 million if that pay was tied to performance. The law ushered in an era of paying CEOs a reasonable base salary (less than a million dollars), and then adding on massive compensation in the form of bonuses and company stock and stock options. The legislation made sense in theory; it seemed like a decent way to tie executive pay to company performance and keep CEO compensation in a reasonable ratio with that of front-line workers. Unfortunately, the execution was often less than evenhanded.

For those who don't understand the inner workings of a publicly held corporation, a CEO's potential pay package is determined by the compensation committee of the company's board of directors. The committee, made up of board members, takes into account the performance of the chief executive (if he or she has been with the company for a while), the salary of others in a similar position, or what it might take in terms of compensation to lure an executive away from another company. This compensation package is then voted on by the board of directors.

Being on a corporate board of directors is indeed nice work. Directors usually receive compensation, which ranges from tens of thousands to hundreds of thousands of dollars a year. (In 2009, board members of the largest 200 companies received an average direct compensation of over $200,000.[23]) In return, they are required to attend half a dozen to a dozen meetings a year and vote on major corporate issues, such as CEO compensation and mergers or acquisitions. Some board members come from within the company — "inside" directors. Others — the "outside" directors — come from outside the company. How are outside directors chosen? Shareholders vote on them from a list of names on the corporate board ballot. How are names selected to be on the ballot? They're chosen by the CEO!

If there were ever a situation rife with the potential for abuse, this is it. First of all, inside directors usually report to the CEO within the company. In other words, they're voting on their boss's pay. If this weren't bad enough, potential outside board members usually run unopposed on the ballot, so if you're on the ballot, you're on the board. In many cases this means that the CEO is hand picking the people who will determine his salary. Outside directors are usually CEOs, or former CEOs, of other companies who attain board membership in a company with the help of that company's CEO. So they're voting on the pay package of someone who helped them boost their own annual income by helping them get the cushy job of sitting on a corporate board. And, in many corporations, the CEO is also the chairman of the board.

And of course, once a few CEOs begin making absurd amounts of income, the whole wage scale for that class of individuals begins to rise. A CEO sitting on the board of another company might work to get that company's CEO an inflated pay package simply hoping that it might increase his own. Even CEOs who haven't packed their respective boards with friends and allies find their salaries rising when directors become convinced they need to pay more to keep a CEO they like.

And board members who might think that a particular pay package is uncalled for have an incentive to vote for it anyway. Once you are on one corporate board, it's not hard to make the ballot for others, unless, of course, you "rock the boat." CEOs talk, and any board member who raises a stink about a CEO's compensation is unlikely to be chosen for any other corporate boards. This has produced a compensation system so out of whack that even CEOs who poorly manage their company make out like bandits, as the following two particularly egregious examples show.

- Carly Fiorina, former CEO of Hewlett Packard, had a combined base salary and targeted bonus of $5.6 million a year. During her reign the value of the company's stock went from $52 a share to $21 a share. While part of the per-share performance is in no doubt due to the busting of the tech bubble, the stock price of Dell (an HP rival) went from $37 to $40 during the same period. Fiorina was responsible for putting over 18,000 employees out of work. Even when the board finally stepped in and forced her to resign for poor performance, she walked away with $21 million in severance pay.[*]
- Robert Nardelli served as the CEO of Home Depot from December 2000 to early 2007. When his tenure started, Home Depot's stock was at $40.75 a share. When he left, six years later, it was at $40.16, even though the company had bought back millions of shares from investors (something that is supposed to drive up the price). Not only was the stock's performance disappointing, but during that time period Home Depot lost significant ground to Lowe's, its chief competitor, whose stock rose 210 percent.[#] Nardelli's ouster probably didn't damage

---

[*] This doesn't include pension benefits, stock options and HP stock holdings, which were estimated at over $21 million. (As if a mini-commentary about the state of our political system and the importance of money over leadership ability, Fiorina ran for the U.S. Senate from California in 2010, garnering 42 percent of the vote.)

[#] In fairness to Nardelli, Home Depot's sales and profits did double during that time period, and he improved the company's return on capital. But this was also the same time frame as the housing bubble, when large sums were being invested in building and remodeling homes, so impressive returns shouldn't have been difficult.

his pride too much, though. He walked away with $210 million in severance, to go along with the more than $240 million he earned while leading the company.

Fully discussing the oversized pay of all corporate chiefs in the last couple of decades would probably require a book in and of itself. But these examples give you an idea of how little compensation has to do with an executive's actual performance.

A 2005 study by Bebchuk and Grinstein showed that during the 1993 to 2003 period, although equity-based compensation (i.e., stock options and stock grants) increased considerably, the growth was not accompanied by a reduction of non-equity compensation (i.e., base salary). The researchers looked at compensation of the top five executive employees at publicly held firms (both "new economy" and "old economy" firms) and found that pay had "grown much beyond the increase that could be explained by changes in firm size, performance and industry classification."[24] In fact, "had the relationship of compensation to size, performance and industry classification remained the same in 2003 as it was in 1993, mean compensation in 2003 would have been only about half of its actual size."* And remember executive salaries had already seen tremendous growth before 1993.

If these demented executive pay schemes weren't enough (and apparently they weren't), there is strong evidence that suggests a large number of companies were cheating the system by back-dating the stock options given to their CEO.[25] While this may or may not be illegal (depending on the legal documents surrounding the Executive Stock Option plan), it certainly violates the intent of compensating an executive, at least in part, with stock options — which is to ensure that the executive's compensation is dependent on gaining a return for the company's shareholders.

What we see when we scrutinize the pay of corporate executives is that a lot of these people have over-inflated salaries, not because of extraordinary performance, but because they have packed the corporate boards with friends and allies and they've

---

* It's also important to note that this study did not consider any increase of executive's pension plans, although these plans regularly consitute a major component of executive compensation.

gamed the system whenever and wherever possible. (To make matters worse, in cases where companies failed and it came time for executives to accept responsibility for their failures, taxpayers were in many cases left with the bill for the cleanup.)

The upper end of the income disparity problem lies in the tremendous growth in salaries for people who fall within the top 1 percent of America's wage earners. If the outrageous incomes of the ultra-rich could reasonably be tied to their performance; if there weren't abnormalities in the market that boost their pay; if there were no forces outside of their own effort that affect their incomes, then there might be a basis for the argument that we should ignore this half of the wage inequity problem, regardless of its effects on the economy. That's not the case. Moreover, the economic effects bolster the contention that this situation should somehow be rectified.

The Cost of Extremely High Incomes
Because high-income individuals tend to have everything they need *and want* (outside of perhaps owning small islands in the South Pacific), most of their extra income goes into savings and investments. Classical economists always believed that this entire amount of savings went into providing businesses with funding needed to purchase productive capital, but we know that this is not the case. If investment funds are directed toward companies that need capital, through venture capital investments or possibly the purchase of IPOs (Initial Public Offerings), then those savings help the economy by providing money for the purchase of plants and equipment. But the bulk of investment funds simply go into purchasing shares of stock in already established companies. These assets don't go toward providing businesses with the funds needed to buy capital equipment; they simply go to whoever owned those shares previously. In other words, they are simply a transfer of wealth from one individual to another. One might argue that the person selling the shares can now spend the money and put it back into circulation, which is true, but that is not what usually happens. People invest in our equity markets as a form of storing wealth. If these savings were simply built up for retirement, our problem would be limited in both time and scope; the money taken out of the economy for retirement savings would eventually begin to re-enter circulation after the investor quit working and began to live off the

store of wealth. But not all savings are spent during retirement, nor are they saved simply for that reason. Some savings are passed from generation to generation or hoarded within massive financial empires, never to be returned to the circular economy. And this overabundance of savings places a drag on economic activity. That slowing is not the only problem created.

As the overabundance of savings (in the hands of a relatively few people) grows, it forms a large wave of money that ends up creating investment bubbles in whatever market becomes its focus. We saw this in the '80s and '90s as income disparity increased and the stock market went on an unsustainable upward climb, similar to the one experienced in the 1920s. After the burst of the tech boom in the early part of the 21st century, the large pool of funds moved on to the housing market. Money not only went into purchasing second and third homes for wealthy individuals, but into mortgage securities that abnormally lowered the cost of home ownership for poorer individuals. This created a massive real estate bubble that eventually took down the economy when it popped. The general economy probably would have collapsed sooner, because the income of average workers wasn't increasing enough to support the demand for America's products and services, but it was sustained temporarily as Americans cashed out the equity in their homes to bolster their incomes. When the housing market deflated, the "refinancing ATM" closed and the economy dropped into the Great Recession. But the investment wave, not nearly as phased as the general economy, has kept right on moving. While unemployment has remained stubbornly high and the economy is still very fragile, equity and commodity markets have done rather well and are both near record territory.

But the economic havoc created by income disparity and large investment bubbles is only the *visible* sign of the trouble created by these outrageous incomes. Behind the scenes lurk other problems.

There is no way to satisfy greed; people who love money and wealth aren't satisfied by having more of it. When the incomes of greedy individuals increase, it simply raises the bar of what they feel is adequate remuneration. Rather than quenching their desire for wealth, extremely high incomes simply distract the focus of these wage earners from the tasks they should be concentrating on, to a focus on how much they are paid. Simply look at the corporate executives and traders who work in our financial system. Their role in our economy is to efficiently and appropriately allocate capital to where it is best rewarded, and to do so in a way that limits risk. But the fact that so many of them were flummoxed when the housing market collapsed shows that they were too focused on their own income instead of on the unstable situation being created in the market.

Anyone who understands economics, particularly the way that these professionals should, could have easily seen that a bubble was forming and that their institutions should have changed the focus of their investments. But they didn't, simply because there was too much money to be made in the over-inflated housing market. The same can be said for workers in the mortgage industry. Their job is to align a potential homeowner with a mortgage that appropriately reflects the borrower's risk and ability to pay. This is not what was happening at the height of the housing boom. Mortgage brokers were making additional money steering borrowers into loans that required little documentation, or had features like adjustable rates and balloon payments, features that obviously some of these borrowers wouldn't be able to manage when they occurred. One might argue that this was a regulatory problem or a side-effect of interest rates being kept too low.

While there is some truth to that, the bottom line is that greed got in the way of professional responsibility.

Let's not blame it all on financial professionals, though. Borrowers have a responsibility as well, to not take on more debt than they can handle. But when the upper echelon of society is continually buying "bigger and better" items — not just homes, but cars and other objects — and this becomes widely visible through the media and popular culture, average individuals will naturally over reach in an attempt to "keep up with the Joneses." Because the income of most workers isn't increasing enough to afford even a fraction of these luxuries, the result is an excess of debt, and in some cases, personal bankruptcies.

Exceedingly high incomes also create other less-seen abnormalities in the economy. Frank and Cook, two authors mentioned earlier, point out that extremely high incomes in winner-take-all type markets create two forms of waste in the economic system. First, the number of contestants attracted to these jobs is larger than it would be if the rewards were more modest. While high incomes attract the best and brightest to these careers, as they should, they also attract a wastefully large number of workers attempting to gain these enormous rewards. As the authors put it, "...the number who forsake productive occupations in traditional markets to compete in winner-take-all markets will be larger than could be justified on traditional cost-benefit grounds." Cities like Los Angeles and New York are teeming with "starving artists" who work menial jobs while they chase their dreams of stardom. Certainly many of these people are talented enough to provide greater benefits to society beyond waiting tables or tending bar, but they take these jobs simply to "get by" until their ever-elusive ships sail in. Society as a whole would be better off if some of the "best and brightest" were drawn to a more diverse array of endeavors — teaching, physical sciences, etc.

The second form of waste that these economists mention is the "unproductive patterns of consumption and investment as contestants vie with one another for top positions." Simply look at the amount of money that colleges and donors to athletic programs spend (sometimes illegally) to win national championships in football or basketball. Obviously it would be better if more of this money went into actually educating college students, but with millions of dollars on the line for a national championship, the total amount invested by all colleges exceeds the amount that is socially optimal. College athletes feel tremendous pressure to take performance-enhancing drugs to elevate their abilities enough to make it into the well-paid ranks of professional athletes. And professional athletes are under the same pressure, knowing that the best player at their position will be paid enough to make the normal salary of a professional athlete look paltry. The overabundant amount of money our society spends on advertising (including during political campaigns) is also due to this "winner-take-all" phenomenon. On a more mundane level, an event like the "boy in the balloon" stunt that took place in Colorado in October of 2009, when police and other community services were used to chase down an air balloon that was supposedly carrying a small boy who had climbed aboard by mistake, is a perfect example of people expending excessive amounts of effort (and social cost) to break into the high-profile world of reality TV.

These problems, created by excessively high incomes, are serious issues that cause a great deal of harm to our society. The good news is that simple steps can be taken to rectify these problems and produce an economy that raises living standards for *all workers*. These steps, presented in Chapters 4 and 5, will not only work, but will require less action and tinkering by our government than we've seen in the last couple of decades.

But before we get into that, let's see what solutions the Republicans and Democrats have suggested.

The Road to Ruin

Because the underlying problem with our economy is an over-inflated level of income inequity, correcting that problem should be the main focus of our solution. It would also be best if we accomplish this feat in a way that requires as little government involvement and expenditure as possible. As we will see, Republicans have failed to even address the problem, and Democrats have taken the wrong approach. And because our chosen leaders have not solved this problem, our economy and society have suffered. What we will see is that each party has actually made our overall situation worse, by trying to rectify our economic troubles without fixing the income predicament first.

Republican Responses to Income Disparity
Republicans have a hard time admitting that income inequity might even be an issue of concern. There might have been a time

when they suggested that the "trickle down" economic policies enacted by Reagan were an attempt to solve the problem. Those policies not only failed, they actually exacerbated our income disparity problem so Republicans usually don't bring it up (even though we tried similar policies at the beginning of this century under George W. Bush). Therefore, conservative solutions to excessive income disparity range from weak to nonexistent.[#] If the subject does come up, Republicans usually give vague responses that dismiss the problem as a side-effect of advancements in technology or differences in the educational level of various workers. Neither argument provides a sufficient explanation.

The "technology" argument claims that as technology has advanced, those who know how to use it have become more productive and thus command higher wages; those who have failed to adapt have been left behind. There are two problems with this line of reasoning. The first is that other nations, such as those in Western Europe, have experienced the same technological advances as the United States and yet have not suffered the same degree of income disparity. If technological advancement were the cause, our hi-tech peers would be experiencing the same degree of trouble that we are. They're not. The second problem is that those who have mastered these new technologies typically aren't the same people earning astronomically high salaries. The highest paid members of our society — professional athletes, high-priced entertainers, and even many CEOs — tend to lack cutting-edge technological expertise. The only exception might be a physician who has specialized in a certain treatment that employs an advanced technology and is therefore earning a million dollars a year because of that expertise. Not only are these cases extremely rare, but they also tend to be in markets that are affected by the winner-take-all phenomenon, so if technology is having an effect, it certainly isn't the whole story.

One area in which technology actually might have had an effect is in the automation of jobs previously done by lower middle-class workers. As automation has increased, the need for a whole

---

[#] In doing research for this segment of the book, I was able to find only one or two articles written by Republicans that even mentioned an income disparity problem, and even then the author did so only in passing.

array of workers has diminished. Go into any automotive plant or other production facility and see how machines and technology have taken the place of many "front-line" workers. Even non-production employees, like bank tellers and service station attendants, have been replaced by simple technologies that allow customers to serve themselves. All of this automation reduces the demand for labor and puts downward pressure on the wages of workers. Republicans don't have a solution for this, other than to suggest that workers need a greater education so that they can take advantage of these technological changes, rather than be hurt by them. (There is at least a partial solution to this problem, and it will be discussed in a later chapter.)

This brings us to the other insufficient explanation: the inadequate education of low-paid workers. Suggesting that a lack of education, or training, is the reason for the falling real wages of most Americans is nonsense as well. Once again, the correlation between those who are highly educated and those who are making disproportionately large incomes is weak. Think of the numerous teachers who have master's degrees, but don't make anywhere near what a CEO (who typically has a master's as well) is paid. Highly paid athletes and entertainers are also rarely on the list of the country's most educated. I don't mean to suggest that education is unimportant or that it can't affect a person's income, but the fact that someone has a college education and is earning a higher wage doesn't mean that she is earning as much as she should. If everyone in America suddenly had a college degree, or even a doctorate, someone would still have to sweep the floors, flip the burgers, and answer the phones. In other words, a higher education may help in individual circumstances, but it's not the solution to the national problem.

Both unemployment and a wage scale that is generally lower than it should be hold down incomes, even for those with a college education. The suggestion that American workers simply need a little more education has a whole generation of workers continually going back to school for a higher degree, or retraining to learn new skills, only to find that the wages in their new career have stagnated almost a much as those in their last. The answer is not an ever-increasing level of education for each worker; it is establishing a system that provides an honest day's pay for an honest day's work, whatever the worker's education level might be.

Human Capital

This raises another key economic concept that should be discussed, though — human capital. Owning physical capital (shovels, backhoes, factories, etc.) isn't the only way to earn more of the revenue pie. Economists have come to recognize what they call "human capital." Just as equipment tends to make workers more productive, education does as well. The more workers know (particularly about the field of work they're in), the more efficient they will be at producing a given good or service. This is why college graduates tend to earn more than those who simply graduated from high school.

But a college education isn't the only way to gain productive knowledge. Every job, from flipping burgers to building nuclear power plants, has a learning curve that workers climb (and through which they become more productive). Theoretically, any time a worker learns a new skill on the job, he or she should be paid a higher wage, because the worker is more productive than a laborer who doesn't have that skill or knowledge. In fact, every time a business acquires a new form of capital for the employees to use, part of the profit derived from that capital should actually go to the worker, who has expanded his or her human capital in learning how to use it. This rarely happens, but it does provide a logical justification for a wider, more equitable distribution of income provided by the use of capital equipment.

Democratic Responses to Income Disparity

In contrast to Republicans, Democrats readily admit that income disparity exists and is an extreme problem. Unfortunately, their remedies tend to create more problems than they fix.

The bulk of solutions we've seen from the Democrats for fixing the problem, at least on the low end of the pay scale, involve big government programs that redistribute income to the poorer members of our society. These include programs like Medicaid, welfare, SCHIP (State Children's Health Insurance Program), food stamps, etc. That so many people have come to rely on these programs as semi-permanent supplements to their incomes rather than as temporary solutions to unforeseen employment problems reveals the problematic nature of such programs. But these programs are really the wrong approach for a number of reasons.

First, they expand the size of government. This expansion usually involves the creation of a large bureaucracy created to administer the poverty-assistance program. While this may create some jobs within the bureaucracy, it usually brings with it an enormous amount of waste and fraud. Medicare fraud, for example, is estimated to cost the federal government (i.e., the taxpayers) over $60 billion a year.[26]

Second, these poverty programs don't tend to discriminate between the three groups of people who usually access them: 1) those who aren't getting by because their wages are too low, 2) those who are willing to work but can't find jobs or are disabled, and 3) those who are simply taking advantage of the system. As we'll see in Chapter 4, there are better ways to remedy the low wage issue for those in the first class. For those in the second group, the unemployed and disabled, America *should* have some set of safety net programs, and with a properly functioning economy the cost of these programs would be surprisingly low. And for those who are able to work, but unwilling, we shouldn't provide anything. Programs that don't make a distinction between these groups can create a disincentive to work and ultimately end up wasting the taxpayers' money.

A recent idea for alleviating the income disparity problem that has earned some traction amongst Democrats is to reduce the contributions that low-income workers make to the Social Security system (i.e. lower the "payroll tax" for the working poor). This would be an illogical step in the wrong direction. While it may help rectify a portion of the income disparity problem, it separates funding of the Social Security program from those most likely to need it. In other words, if low-wage workers are more likely to need the social insurance provided by the Social Security system, they should have to contribute to that system. Otherwise, that program simply becomes another "tax the rich to give to the poor" plan run by our government. The more logical solution to income disparity would be to ensure that workers are being paid a fair wage, and then let them contribute to the government like everyone else.

One tactic used by Democrats to help alleviate the income disparity problem that actually has some merit is the setting of a minimum wage. Given the ever-present existence of unemployment (surplus labor), and a large body of illegal aliens (even more surplus labor), both applying downward pressure on wages, it makes sense to set a floor under wages by creating a "minimum" level that workers must be paid. This can help correct one of the imperfections inherent in the free market system, which assumes that society is always at full employment, and therefore, wages are higher. Some economists would argue that wages should be set through an intersection of supply and demand. That sounds reasonable, but if that market-determined wage is too low to provide for a person's necessities, as it would be with a measurable level of unemployment, then it will create a situation where the government feels compelled to take action in order to mitigate the raft of societal problems that would be a byproduct of low wages.

Regrettably, Democrats have failed to keep the minimum wage at a meaningful level. Part of the problem is that the Democratic Party has never laid out a logical reason for the minimum wage and articulated what living necessities it should cover.[*] Had this been done, a strong argument could have been made for indexing the minimum wage to inflation to make sure that our country's lowest-paid workers can get by without government assistance. As it stands, the minimum wage loses purchasing power to inflation in long periods when it isn't raised and is then bumped in small increments that never fully bring it back to a meaningful level, if it was ever at one to begin with.

EITC – Bipartisanship for a Bad Idea

One supposed cure for income disparity that enjoys bipartisan support is the Earned Income Tax Credit (EITC). According to the IRS, the EITC is "a tax credit for certain people who work and have low wages."[27] It is meant to encourage work by helping to lift the working poor out of poverty. Started by Gerald Ford in 1975, it was expanded in 1986 under Reagan, in 1990 under George H.W. Bush, in 1993 under Clinton, and in 2001 under the second Bush (George W.). Most recently, the EITC was increased for 2009 and 2010 as part of the American Recovery and

---

[*] Today's Democratic Party, in general, seems pretty weak at explaining and defending any of their policy positions in a common sense manner that most Americans can understand. This might account for their inability to gain and maintain power in our government for any extended length of time, even though the financial status of most Americans should make them Democratic loyalists.

Reinvestment Act. In tax year 2009, the cost of the EITC to the federal government was estimated to be almost $58 billion.[28]

In reality, the EITC is merely a subsidy to businesses. Our society, conservatives and liberals alike, have decided that working people should not be living in poverty, and rightfully so. But instead of forcing the business community to pay wages that would keep their workers out of poverty and off government assistance, our government has decided to subsidize their income with the tax money of other workers. Many politicians and economists argue that the EITC provides an offset of Social Security contributions for poorer workers. That follows the same failed logic as the idea of reducing the payroll tax for the working poor mentioned earlier. The better idea would be to establish a ground rule for the economy that forces corporations to pay workers an income they could actually live on, instead of subsidizing them with tax dollars. Then maybe the payroll tax wouldn't seem such a burden.

The EITC does make *some* logical sense from an economic point of view in that it encourages people to work, rather than simply giving them welfare. But higher wages also provide an incentive for people to work, and do so without putting a burden on the federal budget. The EITC is very efficient in terms of getting money into the hands of the poor (i.e., not a lot of extra bureaucracy needed) and its structure provides an incentive for people to earn more, so it might be a reasonable way to help those who can't find full-time employment. But it should never benefit those who are working full time. If our economy has a proper wage structure in place, people will have not only the incentive to work, but the ability to earn an adequate living, even at the lowest end of the pay scale, without having to rely on government handouts.

The Engine Breaks Down and the Fed Steps In

Because the Republicans and Democrats have not properly addressed our increasing level of income disparity, more and more of our country's money has become clustered in one small segment of our society. The fortunate few take home increasingly larger incomes that mostly get squirreled away in investment accounts, where it is unavailable to the rest of the economy. At the same time, the majority of Americans cut back on spending because their paychecks — shrinking in inflation-adjusted terms — don't stretch as far as previously. As this occurs, the demand for our country's products and services falters and our economy goes into recession.

I would be lying if I told you that income disparity is the only catalyst for an economic downturn. It's not. Spikes in the prices of commodities, high interest rates and terrorist attacks are a few of the things that can also cause economic slowdowns and even recessions. But the negative effects of these events would be minimized if our country's income were more evenly distributed. If more Americans had paychecks large enough to afford to set aside savings, there would be a ready reserve families could tap into in times of trouble, rather than wholly relying on the Federal Reserve and the government to bail out the economy. As it is, the average American family lives paycheck to paycheck, has a falling real wage, a negative net worth, and little or no savings. Granted, part of this problem is cultural. Americans need to learn to save more. But you can't save what you don't have. And if the income of the average American family barely covers necessities, there isn't much — if anything — left to save.

So in times of trouble, the Federal Reserve comes to the "rescue" and attempts to revive the economy by expanding the money supply. The initial response to this action is a drop in interest rates, which helps stimulate economic activity. This is analogous to dumping more oil into the sputtering engine; it fixes the problem temporarily by supplying more of the pistons with oil. But increasing the amount of money in circulation doesn't address the heart of the problem. Therefore, it is only effective for a limited time, a reality currently becoming painfully obvious here in the United States. For the last three decades or so the Fed has typically stepped in to provide "extra liquidity" (i.e., more money) every time the country has experienced economic difficulty. But because our leaders have not addressed the real problem — high levels of income disparity — additional currency injections are now proving futile. The amount of money that has been created and pumped into the economy since 2008 is unprecedented. Yet it has done little more than keep us from utter economic collapse.

Adding money to the system also creates an economic problem: inflation. Any time you have an increase in the amount of money chasing a fixed amount of goods, prices will rise. Thus each time the Fed increases the money supply faster than the growth of the economy, we are guaranteed to see inflation.

The inflation, however, isn't always apparent to the casual observer. In fact, there's been a debate raging in the economics and investment community regarding whether the United States is about to see an extended period of inflation or deflation. One side argues that all the additional money that has been pumped into the economy will eventually cause inflation. Those in opposition claim that the economy is so weak that we are bound to see a period of falling prices, similar to what the nation experienced during the Great Depression. I don't think it's out of the realm of possibility to say that both sides are right.

Because one of the remedies to our recent economic hiccups has been to pump more money into the system, we now have an economic engine with too much lubrication. At the same time, we haven't yet corrected the income disparity problem, so too much of this money is pooling in the hands of a relatively small segment of our society. If this extra liquidity causes inflation, and too much of it is residing in the hands of the ultra-rich, wouldn't it make sense that inflation would be most prevalent in the price of things the ultra-rich buy? If you look back over the last few decades, this is exactly what we've seen. Remember in the last chapter when I talked about the huge pool of investment funds moving from investment to investment, driving up prices and causing investment bubbles? That's inflation! The massive run-up in prices that every one is expecting hasn't taken place in the general economy; it's taken place in that segment of the economy where wealthy people put their extra cash: in various forms of investments. And these investment bubbles won't go away until the distribution of wealth and income in our country becomes somewhat more equitable.

At the same time, because the wages of a majority of Americans have stagnated, the general economy is struggling and demand for a whole host of items is depressed. The culmination of all this reduced demand for goods and services in the general economy would suggest that a certain level of deflation is also in order.

So it's possible that both sides are right: those who expect inflation, particularly in assets affected by investment demand, and those who expect deflation, caused by a lack of demand for general goods and services. It's just another sign of how our income disparity level is wreaking havoc on our society and economy.

Budget Deficits and Interest Rates

Of course the President and Congress can't be viewed as sitting around on their duffs during an economic downturn, relying on the Fed to fix everything. So they get into the act as well, with legislation intended to pump life back into the economy. The response usually involves tax cuts or spending programs, or a combination of both. This decrease in government revenue or increase in spending has the unfortunate effect of causing a budget deficit (or increasing the current deficit).

Republican responses to economic slowdowns and recessions tend to focus on what is called the "supply side" of the economy. The theoretical basis behind supply-side economics comes from economist J.B. Say, who wrote that "supply creates its own demand." In other words, as businesses buy machinery and hire workers to produce a good or service, they create demand by putting money into the hands of workers (who are also consumers). Republicans therefore try to enact legislation that favors businesses and capital owners in an attempt to get them to produce more goods and services. The legislation typically involves an assorted package of tax cuts for the nation's wealthy (capital owners), as well as for businesses. The assumption is that the extra money produced by the tax cuts will be used to either start new businesses or expand already existing ones. These cuts usually come in the form of a reduction of the top marginal income tax rate, a cut in the capital gains rate, or a lowering of the business tax rate. The theoretical basis for these cuts makes sense. Unfortunately, the outcome is less than ideal.

If it is true that a cut of the top marginal income tax rate is a "supply-side" tax cut and will, therefore, increase business investment, then we should expect to see greater levels of business investment during periods when the top marginal tax rate is falling and lower levels when the tax rate is rising. The chart below[29] shows that the exact opposite is the case. It compares the top marginal tax rate (measured on the left axis) to private business investment (measured as a percentage of the nation's gross domestic product on the right axis). Notice that following the tremendous cut of the top marginal rate in the 1980s, business investment fell. As that same rate was increased during the '90s, business investment actually grew. If lowering the top marginal rate really improved business investment, these lines would head in opposite directions. Instead, they mirror each other.

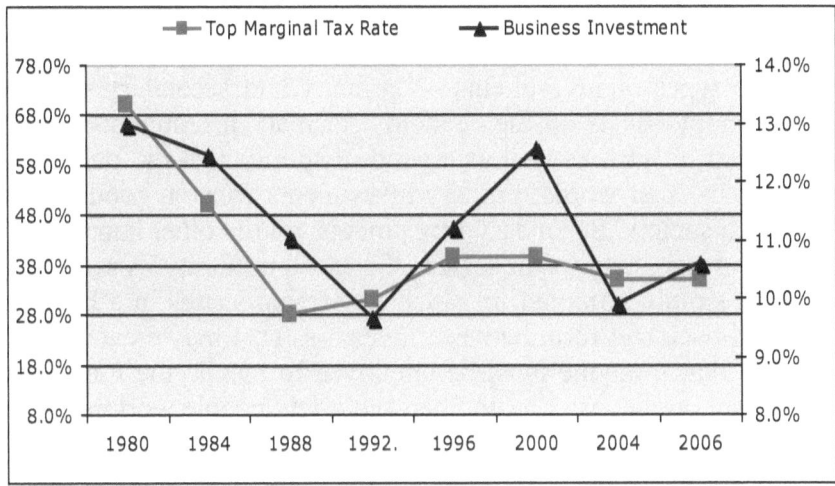

The reason this type of tax cut doesn't work may be the fact that it's not really a supply-side stimulus. Remember, a supply-side stimulus is supposed to encourage businesses to hire workers and invest in equipment. Cuts in the top marginal rate don't do that. They are just tax cuts that favor high-income Americans. What if the people who benefit from these cuts decide to spend the money in a foreign country rather than invest it or spend it in America? We can *assume* that a tax cut for wealthier individuals will result in more business investment, but the facts show otherwise, and I've seen little, if anything, in any of the Republican tax cut bills that would encourage actual business investment.

There is a similar, yet separate problem with our capital gains tax structure. Because capital gains are derived from the increase in value of investment assets, cutting the capital gains tax would seem to encourage these investments, and indeed it does. The problem is that not all investment assets are the same, as I pointed out in the last chapter. Nor do they all have the same impact on our economy. "Primary investments," as I call them, provide the capital that businesses need to build factories, buy machinery, and employ workers — referred to in the chart above as "business investment." Other investments, what I refer to as "secondary investments," are made when someone buys shares in a company on a secondary market, like the NYSE, NASDAQ, or AMEX. Because the money from these investments goes to the individual selling the shares, rather than to the company, it does nothing toward helping the company buy the items necessary for its operation.

If you think about this for a minute, it becomes obvious that these two types of investments — primary and secondary — have very different effects on the economy. Primary investments lead to the creation of jobs, as well as a good or service that the public can purchase. In other words, primary investments create a good deal of value for a society. Secondary investments, on the other hand, create almost no value, no good or service, for the community. A secondary market is simply a market in which ownership shares in a business are traded back and forth between investors. That may create jobs in the sense that someone must be employed to handle the transfer of shares from one investor to another, but a few people working on an exchange such as the NYSE can handle thousands of these transactions a day, so the employment benefits are limited. Primary investments create valuable and productive employment, whereas secondary investments simply transfer wealth. This is not to say that secondary investments and secondary markets aren't necessary, or aren't important — they are — but they're not as important or beneficial as primary investments.

The problem with our capital gains tax system is that it makes no distinction between the two. If someone bought shares in General Electric and experienced a gain of 20 percent over a year's time, that person's gain is treated exactly the same as that of someone who invested in a start-up business and experienced a 20 percent gain over five years, even though the effects of the two investments on the economy are totally different. People find investing in established businesses much easier, and safer, than starting their own businesses or investing in startups, which carry more risk. And because the two are treated the same by the tax system, most investment funds flow into secondary investments, causing large increases in equity markets, but providing little benefit to the real economy. Cutting the capital gains tax only exacerbates this trend. Some sort of distinction needs to be made in the tax system for capital gains that will encourage entrepreneurs to start small businesses while not overly rewarding rich Americans for simply dumping more money into the stock market. Until this is done, cutting the capital gains tax will provide very little economic benefit to the country.

For too long, Republicans have promoted the fallacious argument that rich Americans create small businesses. A little reasoning shows that this doesn't make sense. Are super-star athletes, famous actors and actresses, CEOs of large corporations, and hedge fund managers — the ones earning the outlandish salaries — going to take the time to start a small start-up business? Highly unlikely; these individuals are going to take their extra income and invest it in the shares of established businesses where they can earn a healthy return each year without the high risk and massive time investment that a start-up business entails. Small businesses tend to be started by middle-class individuals in an attempt to escape the dreariness of corporate life and eventually give themselves a higher income.

Therefore, cutting the top marginal tax rate and the capital gains rate do little to actually promote economic growth. Granted, these cuts produce nice gains in the equity markets, but if those gains aren't associated with real economic growth then they will be bubble-like in nature and will ultimately prove fleeting. What these cuts do accomplish, however, is a huge reduction in government revenue, making the budget even harder to balance.

The latest conservative push for tax cuts has been in the realm of the corporate tax rate. The reasoning behind a reduction in taxes on business profits usually follows one of two lines: international competitiveness or job creation. Once again, under scrutiny, both arguments fall flat on their faces.

Rep. Michelle Bachmann (R-MN) promoted the first line of reasoning in her response (for the Tea Party) to the 2011 State of the Union address by claiming that the reason American corporations have been moving our jobs overseas is that our corporate tax rate is too high. If true, this would obviously support her suggestion, and the suggestion of many Republicans, that the corporate tax rates need to be lowered, and doing so would help keep jobs in the United States. The only problem is that corporate tax rates having been falling over the last few decades — the same period of time when we've lost the most jobs to countries like China and India. If the lowering of corporate tax rates has been met with the offshoring of our jobs, should we lower it even more? I'm not suggesting that the tax reductions caused the exporting of American jobs, but that the facts of the situation fly directly in the face of the conservative argument. As if to make the point even more clear, during the 1940s, '50s, and '60s, when our economy was performing at its best, and we weren't losing our jobs to other countries, the top corporate tax rate rarely dipped below 50 percent; it's at 35 percent today. The fact of the matter is: there are other factors which have a much greater effect on whether or not American jobs are exported to other countries.

The other line of reasoning for cutting taxes on business profits stems from a huge misconception — I'll call it the "Joe the Plumber" syndrome — that taxes on a business owner's profits affect how many employees he or she hires. This is complete nonsense. The profits of a business are determined by the demand for its product, the price of that product, and the price of the inputs required in making the product: capital, materials, rent, and labor. Profit taxes have nothing to do with it, because they are determined *after* the profit is determined. Businesses hire labor based on demand for their products and whether or not they can sell the products for more than their cost. If you cut the taxes of a business owner, but that business owner can't sell any more product, do you really think she's going to hire extra workers? Of course not! Alternatively, even if you raise the tax on business profits, a business owner will still hire more workers if she can make and sell more of her product for an increased profit. Any time you hear a politician claim that we need to lower taxes on businesses so they can hire more people, all they're telling you is that they know nothing about how a business works.

This is not to say that corporate taxes should be raised, or that they shouldn't be lowered, only that the Republican reasons for doing so make absolutely no sense. It's as if conservatives are looking for any justification for this policy proposal to hide the fact that they don't have a legitimate one. Ironically, the taxes that *should* factor into a business owner's decision to hire more workers are the ones placed on labor: income and payroll taxes. These taxes directly increase the cost of labor, something that *definitely* factors into the demand for workers. Unfortunately, this is rarely the thrust of Republican tax cuts.

In contrast to Republican responses to economic downturns, Democrats tend to focus on the "demand side" of the economy — putting money into the hands of people who will go out and buy goods and services. This stimulus is typically accomplished through a creation of government spending programs or an increase in current programs. Numerous problems with these outlays exist, though.

The first problem is that these increased expenditures, if not coupled with tax increases, can cause (or increase) a budget deficit. This wouldn't necessarily be a problem if the deficit was short-term in nature, and any debt created was repaid once the economy recovered. But our government has proven time and again that it has a difficult time balancing the budget, let alone running a surplus. Deficit spending can also increase interest rates and crowd out productive investment unless it is matched with an associated increase in the money supply, a measure that causes its own problems, as noted above.

Increasing government spending also amplifies the problems that already surround these outlays. Any time the government spends money, it is directing the economic resources of the country. While it is necessary and beneficial for the government to command *some* of these resources for various community benefits (roads, schools, national defense), spending decisions made by the state tend to be less efficient than those made by the market. Therefore, decisions regarding government expenditures should be on the basis of communal needs of the country and the people's willingness to sacrifice part of their earnings, in the form of taxes, to pay for those benefits. Suggesting that a sudden increase in government spending is necessary to stimulate the economy throws a wrench of urgency into a decision process that should be more deliberate.

Federal spending also tends to be a target of fraud and abuse. Whether it takes the form of companies padding their profits with lucrative government contracts, lobbyists convincing politicians to spend money on projects of no significant value, or prisoners trying to take advantage of first-time home buyer tax credits; it's obvious that when the government starts handing out money, the pigs come to the trough to feed. While limiting government spending may not necessarily fix these abuses, the more we spend, the more money we are likely to lose to those who care more about their own financial benefit than the fiscal health of their country.

The timing of government stimulus expenditures is also problematic. Congress could be considered one of the slowest moving animals on the Earth, and getting our elected representatives to pass a stimulus bill, even in the face of a recession, takes time (often due to partisan bickering and stall tactics). Even if the government could enact legislation faster, getting the money out the door and into the hands of the public doesn't happen overnight.* What this means is that usually the economy has started to recover by the time the stimulus becomes available.

Jobs produced or supported by federal stimuli also must be paid for once the stimulus is cut off, or those jobs could be lost. For instance, if a Democratic stimulus bill includes funding to put additional police officers on the street, what happens when the stimulus funding ends? Those cops must be laid off, or the states where they work must come up with the tax revenue to pay them. It's assumed that the economy will have picked up enough that the additional state revenue will be there, but that isn't always the case.

Democrats, in an attempt to show that they, too, support tax cuts, have also included "targeted tax cuts" in their stimulus measures. These tax cuts usually aren't a decrease in any particular marginal tax bracket, but provide a tax deduction or credit for spending on certain items, such as college tuition or fuel-efficient technologies. This latest round of stimuli included tax credits for "first-time" homebuyers and a "cash for clunkers" program intended to get gas-guzzling vehicles off the road.

Targeted tax cuts are good in the sense that they reward taxpayers for making purchases that have an external benefit to society. But in many cases the incentive to make these expenditures already exists in the economy. The fact that college graduates earn more than those without a college degree is incentive enough for people to want to attain a college education. If Americans aren't attending college, or would like to but can't afford it, this is a problem caused by income disparity, not a problem requiring a government incentive program. In cases where the appropriate incentives don't exist, there are ways to provide those incentives that make more sense from a market point of view and are better for the federal budget, as we'll see in Chapter 6.

Targeted tax cuts can create additional problems by cannibalizing consumer spending of future years. If someone plans

---

* And the faster the money is doled out, the less control exists over where it goes (i.e., more fraud and abuse).

to purchase a furnace next year, but buys a fuel-efficient one now to take advantage of a tax credit that won't be available next year, that might stimulate the economy now, but it comes at the expense of spending that would have been done the following year. Besides, targeted cuts further complicate what is already an extremely complex tax system. As is the case with most Democratic solutions: the goal is noble, but the technique is flawed.

Conclusion

Our government has a constitutional responsibility to "promote the general welfare" of the nation. The present state of affairs in America shows that our leaders have failed miserably in this regard — our economy can't get off its deathbed, our federal budget seems impossible to balance, and our dollar continues to fall in value. And that just describes the situation in broad strokes. Jobs are increasingly harder to find; ones that can be found don't pay well; raising a family for most Americans requires two incomes *and* credit card debt; crime, both white collar and street crime, are rampant; and even if you manage to avoid all of these complications and provide a decent life for your loved ones, your acquired assets might disappear in the busting of the latest investment bubble.

This failure of our government stems from the fact that neither Republicans nor Democrats have developed the right response to the underlying problem in our society: our real-world economy doesn't match the theoretical one on which it is based, and that creates extreme levels of income disparity. Republicans have tried to ignore this reality, and have even enacted policies that make the situation worse. Democrats have recognized the problem, but tried to fix it by "providing" the general welfare instead of "promoting" it — undermining the strengths of our economic system, despite its imperfections. It is a combination of these flawed responses that has gotten the United States into the predicament we're in today. Republicans have pushed tax cut after tax cut, crimping government revenue while providing little actual economic benefit. What these tax cuts did provide was the cash needed for casino-style bets in successive investment bubbles, the bursting of which threatened the whole economy. At the same time, Democrats have created program after program aimed at reducing the inequities in our society and reviving economic demand. While they have failed at attaining either of these goals, they have managed to expand the size and cost of our government and create some nice disincentives to work. And, all the while, the Fed has gotten into the act by expanding the money supply and ignoring its oversight role of the banks in our financial system. This reduced the value of the dollar, kept interest rates dangerously low, and provided the environment needed for the investment casino mentioned earlier.

As we'll see in the next few chapters, what is really needed are some simple fixes — ground rules essentially — that address the imperfections of the free market system. These "ground rules" aren't complicated or over-burdensome, particularly when compared with the regulations, government programs, and tax tinkering the Democrats and Republicans have proposed and enacted, but they will promote the general welfare. These ground rules will improve the economy and provide a foundation, both rationally and economically, from which to reduce the size and scope of government. Neither goal looks achievable with the solutions proposed by the parties presently running the country.

A New Capitalism

Given the problems pointed out in the first chapter, it's obvious that the growing levels of income disparity we've been experiencing in the United States can't continue. The social and cultural ramifications are bad enough, but the economic consequence has been a downturn so severe that it elicits comparisons to the Great Depression. As we saw in the second chapter, this problem is born of the differences between the world we live in and the theoretical world upon which our economic system is based. This doesn't mean that we should abandon free market capitalism, but that we should be its master rather than its slave. Ignoring the problem — as Republicans have done — or trying to solve it through a constant expansion of government — as Democrats have done — will obviously not work. That should be apparent from the faults in these policies, pointed out in Chapter 3, if not from the last 40 years of our country's history. We've been trying different variations of each party's ideas for decades and our problems are getting worse, not better.

What is needed is a different approach. In this chapter, and the following two, I put forth a plan that will rectify our economic problems and lay the groundwork for balancing the federal budget and paying down our national debt. While that sounds good, the benefits aren't limited to the realms of economics and finance. This proposal will also put us on the road to energy independence and help us limit the damage we do to our environment. I can't stress enough that the policies put forth in the rest of this book are proposed because they are necessary for rectifying our country's problems, not because they help fulfill some liberal or conservative political agenda. As I mentioned in the Preface, this is *not* an ala carte menu. It is a comprehensive economic and fiscal proposal. As such, each part is a necessary component in the overall plan. We can't simply enact the suggestions we like, and ignore the ones we don't.

If our government makes these changes, and faithfully maintains the general purpose of them, there'll be no need for massive bailouts of Wall Street and big banks; less need for environmental regulation; no need for subsidies to promote energy efficiency or new clean technologies; and programs like Social Security, Medicare and Medicaid will be much less essential. Most importantly, the economy will operate in a fairer, less volatile manner, without the need for all the government tinkering we've seen in the last few decades.

Let us begin…

A Proper Minimum Wage

Because our country is not at full employment, as economic theory assumes, the wages that we see in the marketplace for most Americans are lower than they otherwise would be. Therefore, the first step we should take is to establish an appropriate minimum wage. Before I set a monetary value on what that minimum wage should be, it makes sense to establish a standard for what someone working at the minimum should be able to afford.

I would hope that all Americans could agree that anyone willing to work a full workweek in the United States should be able to afford basic necessities and get by without government assistance. What's the point of working if you still have to be on the dole? If working barely provides a better living than not working, where is the incentive to work?

Many progressive groups refer to a "living wage" and suggest that this should be the model for a minimum wage because it is calculated as a person's minimum necessities for living. Most living wage models include seven components, or needs, that must be covered in order for compensation to qualify as a living wage: food, shelter, health care, child care, clothing/other, transportation and taxes. The minimum wage that I suggest here includes many of these items, but does make some fundamental changes.

It's reasonable to suggest that the minimum wage should provide one the means to pay for shelter, modest clothing, food, and funds for getting to and from a place of employment. In terms of shelter, a minimum wage should certainly allow people enough of a rent payment that they don't have to live in squalor and potentially enough of a shelter allowance that if they acted wisely in the market, they might be able to afford ownership of a small home. In terms of clothing, it would be ludicrous to suggest that someone making the minimum wage should be able to afford the finest fashions, but it's not unreasonable to think that this person should be able to afford clothing suitable for his or her employment, and nice enough to provide a good representation if interviewing for a better job or career. As with the clothing and shelter segments of the minimum wage, the food and transportation segments should provide for modest expenditures in these areas. Minimum wage workers shouldn't be able to afford eating filet mignon every night, but they also shouldn't be relegated to surviving on Top Ramen.® A healthy, balanced, yet modest diet seems like a reasonable floor for any worker in the wealthiest country in the world. Likewise, a minimum wage worker shouldn't be driving around in a Porsche or Ferrari, but the wage should be enough to provide for modest, sensible, worker-owned transportation.

For those of a liberal bent who feel this may not be enough, remember: this is the *minimum* wage. For those who are more conservative and feel that what I'm describing might be too generous, remember: we're not talking about welfare recipients, we're talking about people who are working. The minimum wage should be enough to provide a reward for working and give workers the opportunity to "pull themselves up by their bootstraps."

It's also only right that working people be able to afford health care. Health insurance could be provided by the employer, or the employers could provide an addition to the minimum wage so that workers could buy their own insurance. Ideally, this would be a basic health insurance package that would prevent bankruptcy in the event of a medical emergency, but isn't so generous that it encourages unhealthy habits like smoking or excessive drinking. Ultimately, the type of insurance package available and the habits it promotes or discourages is a matter for the healthcare market. For the present discussion, suffice it to say that it only makes sense that people who are working should be able to keep themselves healthy enough to continue being productive.

All Americans should pay taxes. We all benefit from various public services and we should all contribute, to some degree, to funding those services. Therefore, the minimum wage should provide enough money so that the employee can pay local, state, and federal taxes (to be covered later).

When calculating a living wage, most economists include an allotment for child care. I, however, disagree with its inclusion. While the minimum wage should be enough to provide a laborer sufficient income to cover the needs of a spouse, it shouldn't be enough to cover the costs that go along with having children. We're talking about a *minimum* income level. And while this income level should be high enough to keep couples out of poverty, there needs to be some limit as to what this "minimum" will cover. Those who are stuck working for the lowest wages in our society need to be able to recognize their situation, and make choices appropriate to that lot. Having numerous children, or even an initial child, and then proposing that the minimum of earnings within a society support that choice seems overly generous. If we're asking businesses to sacrifice a higher wage than might otherwise be seen in the open market, we should also ask workers benefiting from this higher base wage to sacrifice some of their personal choices to make that salary work.

While I don't include an allotment for child care in the minimum wage, I do feel there should be an allotment for "investment." If the way to prosper in a capitalistic society is to own capital, everyone should have that opportunity for prosperity. If we pride ourselves on being a nation where anyone can prosper, shouldn't the minimum wage of our nation also include a small portion devoted to that end? An investment allotment within the minimum wage also makes sense so that workers are able to save for their own retirement and not have to rely on the federal government to take care of them once they stop working. It's also worth mentioning that this "investment" portion wouldn't necessarily have to go to traditional equity investments. A minimum wage worker could save this portion for starting his/her own small business. Or this portion could be saved and used toward gaining a higher education — an investment in human capital. Once again, if we expect our fellow countrymen to improve their own lot without relying on government help, shouldn't the *minimum* of earnings provide them the means to do so?

The following is a layout of the various segments of an appropriate minimum wage and the values that might be associated with those segments on an annual basis.

- Housing - $8,400 ($700/month)
- Food - $2,400 ($200/month)
- Transportation - $3,600 ($300/month)
- Clothing/other - $700
- Health insurance - $5,040 ($420/month for a couple)
- Investment - $3,000 (Roughly 12 percent of gross income)
- Taxes - $2,000 (assuming $1,500 for federal, $500 for state and local)

Based on a 2,080-hour work year (40 hours a week, 52 weeks a year), this translates into an hourly wage of just over $12 an hour.

Now, it's obvious that prices in some areas of the country are higher than those in others. For someone in Manhattan, where the average apartment runs $3,800 a month, the housing allotment may seem too low; for someone in Boise, Idaho, it might seem slightly generous. Some people may feel the allotment for health care is somewhat low; others, too high. While the exact figures might be up for debate, what I've tried to do is find a reasonable nationwide average for each allocation. Whatever the case, I would hope we could all agree that what I have laid forth here isn't grossly unreasonable. It's not overly generous, as liberals might want, or overly frugal, as conservatives might like, but represents the basic resources an American couple would need in order to take care of themselves without the government's help. Even if you think minor tweaks should be made to the values I've used, it should still be painfully obvious that our minimum wage needs an upward adjustment of somewhere in the ballpark of five dollars an hour (presently, the minimum wage is $7.25 an hour). Even if the employer provides health insurance for the employee and spouse —

a rarity — and therefore that segment is removed, the necessary hourly wage is still over $9.65 an hour — almost two and a half dollars higher than the present minimum. One could also easily argue that since the minimum wage has been so low for so long, workers on the lower end of the scale have built up debt simply trying to get by and should be over-compensated for a period of time to help offset their indebtedness.

As I've noted before, if there were no unemployment, as the theory underlying free market capitalism assumes, wages would be higher across the spectrum. This however, doesn't give us a definitive amount of what the lowest of those wages would be. However, it's reasonable to assume that if workers were guaranteed work somewhere, because labor demand perfectly matched supply, the least that workers would demand would be coverage of their minimum needs. Once again, if working for a living doesn't cover your necessities, why would one engage in it? The only reason that workers labor for less than that amount now is because there is an overabundant supply of workers in the labor market. The only thing this adjustment in the minimum wage would do is replicate the wage conditions we would see if the real world market matched the theoretical one. And if the economy is able to function properly in the theoretical world with a higher wage rate, then it should be able to do so in the real world as well.

If you think that a minimum wage north of ten dollars sounds extreme, consider this: a lot of Americans would consider twelve dollars an hour a good wage, but it's barely enough to cover their necessities! That's one of the reasons why so many people rely on some sort of government assistance during part of their lives, even if it's just basic entitlement programs like Medicare and Social Security. The low end of the wage scale is simply not high enough for self-sufficiency.

Now, of course, some minimum wage workers will take these extra earnings and use them wisely, some will not. Some will take the opportunity that a realistic minimum wage should provide and save some money to invest in themselves and their future. Others will squander it on lottery tickets or dig themselves into a hole by having more children than their income can afford. This should not be the government's concern. All the government should be concerned with is setting fair ground rules so that those who are willing to work can provide for themselves, not making sure everyone is shielded from the mistakes of their own imprudence. Americans should be able to say that the lowliest of workers in our country has the opportunity to pick himself up and make something of himself. A reasonable minimum wage is the first step toward that goal.

The next question becomes "how should this adjustment be made?" Large, sudden changes in the minimum wage can cause huge economic disruptions in the short term. Therefore, while recognizing the need and justification for a higher minimum wage, we should also give deference to the stability of the overall economy in the short term. I recommend raising the minimum wage by 75 cents to a dollar per year for the next five or six years, so that it crosses the twelve dollar threshold by the end of the adjustment period. These increases will go a long way toward bringing the minimum wage up to a realistic, logical level without unduly disrupting the economy. Due to inflation, a recalculation of what the minimum wage should be would need to be done at the end of the adjustment period to determine future increases, if necessary.

Opponents of the minimum wage seem to have an endless stream of objections to its existence and any increase in its value. Some of these even directly contradict each other. For instance, some opponents will claim that the minimum wage causes inflation, and that so few workers make the minimum wage that raising it won't do much good. If either of these were true, it would obviously cancel out the other. If so few people make the minimum wage that raising it wouldn't do much good, then how could it cause so much inflation? Regardless of the lack of a coherent logic behind some of the objections, it is worth giving them consideration.

The first objection centers around the perceived inflationary effects of a higher minimum wage. Let me begin by saying: *It is completely possible that a higher minimum wage will cause inflation.* However, it is more difficult to establish a causal relationship between an increase in the minimum wage and inflation than some politicians and pundits would have you believe. The last few decades provide a perfect example. The minimum wage sat stagnant at $3.35 an hour for most of the 1980s, while inflation for that decade ('81-'90) averaged 4.7 percent. This inflation level might be skewed somewhat by the high inflation levels of the early '80s, but even when we look at the second half of that decade, we find an average inflation rate of almost 4 percent. In contrast, the minimum wage was increased four times during the 1990s (going from $3.35 in 1989 to $5.15 in 1997) and the average inflation rate for that decade was a much lower 2.8 percent.

As Nobel Prize-winning economist Milton Friedman put it, "Inflation is always and everywhere a monetary phenomenon." In other words, as we saw in Chapter 2, as long as the government allows the money supply to grow at a faster rate than the economy, the country will experience inflation. Indeed, during the latest recession, both Republican and Democratic administrations, in conjunction with the Fed, have expanded the money supply by historic amounts. This has brought numerous predictions for much higher levels of inflation, with little or no mention of increases in the minimum wage.

But is it possible for the wages of workers to be raised without this causing the prices of the goods they produce to increase? Absolutely. Higher income levels don't have to be paid for through higher prices. These income increases could be produced by lowering the wages at the higher end of the company. For instance, the CEO and other executives could reduce their pay, and hence have the overall wages of the company stay the same.* Increases in the wages of front line workers could also come from a lowering of company profits, without an effect on prices for consumers. While some may object to the fairness of these alternatives, it's obvious that higher prices aren't the only source for producing higher wages for workers. (It should be noted, though, that lower salaries for corporate executives and lower corporate profits would both tend to help reduce the income disparity problem that needs to be solved.)

Debate has also raged in the economic community about the effects of minimum wage increases on workers (particularly unskilled workers) and small businesses; the thought here being that increasing labor costs might reduce the demand for labor. Recent studies seem to conclude that minimum wage increases benefit the income of lower wage workers without affecting the number of workers employed.[30] In other words, raising the minimum wage doesn't cause owners of small businesses to reduce the number of workers they have on staff. But while they're not terminating low

---

* This, of course, begs the question: "Why don't pundits and politicians complain about the inflationary effects of increases to CEO pay packages and bonuses?"

wage employees, can it be said that they're seeing a benefit? Well, look at the clientele of most small businesses. Is it made up of the top 1 percent of wage earners (whose wages have been increasing for the last four decades), or do small businesses usually get their patronage from shoppers on the lower ends of the income scale (who have actually seen their purchasing power stagnate, or fall, since the '70s)? While some small businesses do indeed cater to the elite (in trendy sections of major urban areas), most get their revenue from working class people who would have more money in their pockets if wages on the lower end of the income scale were raised.

During my college years, I worked at a little "mom and pop" grocery store. The owner of the store, Jerry, always made sure to pay a little more than the minimum wage. Jerry hated increases in the minimum wage. He saw minimum wage hikes as higher labor costs. But what he failed to realize is what most conservatives fail to realize when it comes to a discussion of the minimum wage: increases in wages not only affect costs, they affect revenue. You see, most of the people who came into Jerry's store were on the lower end of America's pay scale. Each increase in the minimum wage meant that most of his customers would probably be walking in the door with more spending money. In fact, one might argue that increases in the minimum wage were critical for the survival of Jerry's Market. Like a lot of small businesses, Jerry's competition tended to be large chain stores — grocery stores like Albertsons. These large, "big-box" retailers can get away with charging lower prices than their small business competitors. Thus if wages remain stagnant while prices increase (and we've seen that they increase whether wages are being increased or not), consumers are more likely to start shopping at large corporate chain stores. Why do you think stores like Wal-Mart have become so popular? People don't shop at Wal-Mart because it's a pleasurable experience, or because they personally know the owner; they shop at Wal-Mart because that's all they can afford. *Increasing the wages of the working poor and lower middle class allows them to patronize small, family-owned businesses*, instead of forcing them to shop at "big-box" chain stores where prices might be lower. This might be the most important point in the whole minimum wage debate, although it's usually completely missed by most politicians and pundits.

Some economists and politicians suggest that raising the minimum wage will have little if any effect on the economy because so few workers are currently earning the minimum. But the fact that so few workers are earning the minimum wage may be a reflection of the fact that it has deteriorated to such a degree in real value terms. Moreover, assuming that only minimum wage workers would be affected by its increase assumes that the rest of America's employed population is earning well above the minimum. This is simply a fallacy. While only 3.6 million Americans earned the minimum wage or less in 2009,[31] roughly 32 million earned less than $12 an hour,[32] one quarter of the U.S. workforce.[+] All of these workers would benefit from the minimum wage increase I proposed above. Additionally, one could hardly expect that someone earning $16 an hour today would continue to accept that wage once the minimum has been raised to over $12 an hour. It's reasonable to believe that an increase in the minimum wage of this size would lift the earnings of over 67 million people — 50 percent of the workforce. This would be a major step in getting the "oil" in our economic engine circulating to all of its cylinders! And it's not like the money isn't out there. Corporate profits are near an all-time high and corporate executives are certainly managing to "feather their nests" to a greater degree every year.

Because capitalism will always tend to pool money in the hands of the few, and drive down the wages of the many, the greatest thing a government can do to stabilize the economy in the long run is to establish a reasonable floor for wages. But in reality our government has never done that. Despite its existence for three quarters of a century in this country, our minimum wage has never been set at a level that would provide the very benefits workers would easily demand if the qualities of the theoretical free market existed in reality. This plan does that.

Income Tax Changes

Since it will take time to get the minimum wage up to where it should be — as well as see adjustments in wages above the minimum wage — we must find other ways to alleviate the income disparity problem. One of the ways to do this is to make changes in

---

[+] Once again, the fact that roughly one quarter of *working* Americans aren't making enough to take proper care of their financial needs should be a strong wake-up call that the American wage structure is out of balance.

the income tax structure. If done properly, modifying and simplifying the income tax system can not only help alleviate income disparity, but eliminate many of the negative economic impacts caused by its very existence.

The taxing of income is, for the most part, a counterproductive idea. Any time an item or activity is taxed its final price increases and the demand for it naturally falls. Work is no different. Taxing wages increases the cost of labor to employers and reduces their demand for workers.* Unfortunately, our government is highly dependent on income taxes for its revenue and debt payments, so complete elimination of the income tax would prove difficult.

Another effect of taxing income is that increasing a person's marginal tax rate reduces that person's incentive to work more. In other words, when people know that the next $1,000 they earn will be taxed at a higher rate (resulting in a smaller amount taken home than from the previous $1000 earned), they are less likely to put in the effort required to earn that income. This effect isn't as great, however, at the polar ends of the income scale. At the lowest end of the income ladder, laborers are typically so cash-strapped that they are more concerned with overall take-home income than they are with the marginal net income produced from each additional unit of effort exerted. At the highest end of the wage scale, income is much less correlated with effort — as we saw in Chapter 2 — and thus higher marginal rates can reduce take-home income without having a meaningful effect on exertion. Simply consider the compensation of Dick Fuld, CEO of Lehman Brothers, who was paid $500 million in 2007. In 2008, his firm declared bankruptcy, yet you'd be hard pressed to find anyone who believes that decreasing his taxes by a million dollars would have provided the incentive for him to work harder to keep his firm solvent. At that range in the pay scale, little relation exists between one's remuneration and one's incentive to work.

---

* This would seem to be a direct contradiction to what I said in the section above about the minimum wage, but it is not. Increases in labor costs that come from *wage increases* go directly into the hands of workers, who are also *consumers*, thus keeping the money in the local economy and creating demand for the same businesses who pay those wages. Increases in labor costs that come from income taxation go into the vacuum of government, and who knows where they'll come out.

Besides the negative effects of taxing earnings, our income tax system has become a complicated labyrinth of rates, deductions, credits, exemptions and eligibility phase-outs. Instead of adding more layers of complexity to the system, it's time to start over with a system that is simpler and more efficient.

Taking these factors into account, the adjustments we make to the income tax system should help reduce income disparity, eliminate taxes on as much of the labor force as possible and greatly simplify the tax system. It would also be helpful, given the state of the economy, if the modifications produce a stimulative effect. The following changes do just that:

1. Eliminate the income tax on the first $100,000 of taxable income earned by a single filer ($150,000 for a couple filing jointly).
2. Establish two income tax rates. The first marginal rate would be applicable to taxable income from $100,000 to $900,000 (for single filers) and would be a rate of 40 percent. The second rate, of 85 percent, would apply to all taxable income over $900,000 for single filers ($975,000 for married filing jointly).
3. Capital gains should be taxed as normal income. The only exception to this would be increases gained through the disposition of businesses where the seller was part of the initial investment group. These "IPO" gains would be taxed at 15 percent (unless the seller pays a lower rate on his/her normal income, in which case the gains would be taxed as normal income).
4. Eliminate all exemptions — for filers, spouses, children and other dependents (Line 42 on Form 1040)
5. Eliminate the following deductions:
    - Educator expenses (Line 23 on Form 1040)
    - Certain business expenses of reservists, performing artists, and fee-basis government officials (Line 24 on Form 1040)
    - Moving expenses (Line 26 on Form 1040)
    - Penalty on early withdrawal of savings (Line 30 on Form 1040)
    - Tuition and fees deduction (Line 34 on Form 1040)

- Domestic production activities deduction (Line 35 on Form 1040)

The system outlined here is a lot simpler than the income tax system we have now. Step number 1 completely eliminates the senseless burden of income taxes for roughly 95 percent of tax filers. Steps 1 and 2 provide a huge tax cut for the largest portion of wager earners in America (particularly in the middle-class). In fact, compared with the 2010 income tax system, almost everyone earning $500,000 a year or less would get some level of income tax relief (single filers with an income of $500,000 would see a slight increase). This should provide an enormous stimulus for the economy, not only by putting more money into the hands of people who will spend it, but by providing middle-class wage earners the seed money to start that small business of which some of them have always dreamed. In fact, for small business owners, this tax reduction would help offset the higher wages they would have to pay due to the higher minimum wage until the economy recovers enough to boost their revenue. In other words, the profits of small businesses might suffer slightly the first year after the plan is implemented, but the take-home pay of small business owners would be less affected because of the decrease in taxes. Additionally, the very high threshold at which the second marginal rate kicks in allows present small business owners to earn a healthy, reasonable return on the time and investment they've dedicated to their venture.

While the first marginal tax bracket is comparable to the rates of our current code, the second marginal bracket is higher than we've seen in this country for over thirty years and merits some discussion. Republicans will claim that the tax increase on the extremely wealthy caused by the second rate is unfair, will kill the economy, and will reduce the incentive to work for those who fall in that bracket. This is simply not the case.

As was explained in the second chapter, most of the money made at the highest end of the pay scale isn't earned; it comes from imperfections in the market or plain old manipulation and unscrupulous behavior. But this begs a question: at what point does the contribution to earnings from a worker's effort or intelligence get overwhelmed by these abnormalities and acts of greed? Is it $100,000 a year? $500,000 a year? $1 million? It's not an easy question to answer. For the purposes of this plan, that amount was assumed to be in the $900,000 to $1 million range; that is where the hefty 85 percent marginal tax rate takes effect. Up to that point, workers are able to keep most, if not all, of their income. Even with the huge increase in the top marginal rate, most families making up to $500,000 would still see a reduction in taxes compared with the present system (depending on filing status, level of itemized deductions in the current system, etc.). Given the circumstances surrounding the income earned at the near-million-dollar-a-year level and above, one could hardly say that the above income tax proposal is "unfair."

As mentioned, our country has seen a top marginal tax rate in this range before. (During the Presidency of Dwight Eisenhower, a Republican, the top rate was at 90 percent.) Was the economy associated with that period of high marginal rates the economic disaster that modern-day conservatives would have you believe? Hardly. In fact, it was one of the best economic periods our country has ever seen. During most of the 1940s, '50s, and '60s, when our economy was at its best, the top marginal rate was at 80 percent or higher, so the argument that this rate will hurt the economy is simply nonsense. The historical evidence alone refutes this argument. Moreover, as we saw in the previous chapter, increases in business investment have been associated with increases in the top marginal rate, so this tax increase might actually *improve* business investment.

But the benefits don't stop there. Having a top marginal income tax rate this high will not only deter greed, it will limit the influence that a very small percentage of Americans currently have over our government. Because wealthy Americans can contribute so much to political campaigns, they have a much stronger say in the direction and policies of the country. This tax system will help bring us back to a more fair and democratic environment. Reducing the amount of after-tax income at the top of the pay scale will also reduce some of the froth that we've been seeing in our investment markets, making the economy more stable and sustainable.

Some will argue that this high marginal rate will create a greater incentive for the wealthy to cheat on their taxes. But because so few people will have to pay the income tax — and most of the ones who do pay will actually see a reduction in what they owe — the IRS will have a smaller pool from which to audit. This should increase the chance that one of these taxpayers will face an inspection of their filing and reduce the incentive to commit tax evasion.

Finally, let's consider the argument that this increase in taxes for the limited few will discourage them from working and producing. I pointed out earlier that this isn't applicable at the upper end of the pay scale, but in an economy with underutilized resources, one could claim that the argument isn't applicable *at all*. You see, a loss of production caused by higher taxes only matters if all workers and resources are fully utilized. If this is the case, no one can step in to provide a substitute for the discouraged worker's production. But in a society with high unemployment and underutilized capacity, this isn't really a problem; if a worker decides to quit working halfway through the year because she doesn't want to face the higher marginal tax rate, someone should be willing to step in and fill her space — particularly if her income is high enough that she would face that top rate. Granted, this argument is highly theoretical — more factors are involved than worker utilization — but that never seems to stop conservative politicians and pundits from using these types of arguments to promote their agenda. The bottom line is this: if some income earners want to quit working because of this top marginal rate, there are plenty of able-bodied, able-minded workers who would gladly take their place.

Step 3, an adjustment to the treatment of capital gains, puts the world of investing in its proper place, without massive government legislation or additional new bureaucracy. Billionaire investor Warren Buffett pointed out a few years ago that he pays a lower tax rate than his receptionist, because the bulk of his income is derived from capital gains. The case is the same for many highly paid hedge fund managers — some of whom make a billion dollars a year — who receive a lot of their remuneration in the form of gains on capital. Having investors — particularly those who may not necessarily be providing productive capital for businesses — pay a lower tax rate than workers doesn't make sense. This change should also counteract some of the greed running rampant on Wall Street and threatening the American economy by limiting the lengths to which some investors will go in order to produce colossal returns. While these particular risk takers have the incentive of individual gains, the potential costs have increasingly been paid for by society at large, either as a matter of policy or through their general economic consequences. If investors know that it's almost useless to make more than $900,000 in capital gains (because they'll be giving 85 percent of it to Uncle Sam), they'll be less likely to place highly leveraged bets in the market in hopes of getting massive returns. Instead of passing thousands of pages of legislation in order to curb the shenanigans on Wall Street, why not simply eliminate them by reducing the incentive to engage in them in the first place?

Most importantly, Step 3 provides a lower tax rate for gains made on IPO-type investments. This will do two things. First, since the initial investors — whether the entrepreneur, venture capitalists, or investors in an IPO — will pay no more than 15 percent in taxes when they dispose of their holdings in a business, more investors will try to focus their investment funds toward enterprises where they can "get in on the ground floor." This will drive financial capital to where it is needed most: into helping small businesses get started. The abundance of funds looking for IPO-type investments should greatly reduce the cost of capital for start-up firms. Second, it will encourage investors to hold on to these initial investments for a longer period of time. If an investor has a nice gain on his initial investment that helped Tommy's Bakery grow and expand, why would he sell that investment to buy shares in IBM (an investment where his capital gain would be treated as normal income)? The IBM investment would have to grow at a much faster rate in order to give him the same, after-tax return. This will not only provide a major funding boost to small businesses, but also lengthen the investment time horizon of investors, which would reduce volatility in the markets as well.

With the lower tax brackets removed, a lot of the deductions that have been added to the tax code over the years simply have no function, or make little sense. They are therefore removed (steps 4 and 5). This greatly simplifies the tax code and the process of filing one's taxes.

Best of all (at least from the government's perspective), this proposed tax system would bring in roughly 17 percent more revenue than the current income tax system, by my estimation. This means that all the benefits mentioned above can be gained without an associated increase in the deficit.

Estate Tax

Increasing the minimum wage and restructuring the income tax system will go a long way toward rectifying current and future income disparity levels. But because we've been living with an ever-increasing level of income disparity for almost four decades, a lot of wealth has accumulated in a relatively small number of hands. Simply look at the chart below, illustrating the distribution of ownership of the nation's assets and liabilities in 2007.[33] The only thing that seems reasonably distributed is personal debt!

Table 4: Percent of Total Assets Held by Wealth Class, 2007				
Asset Type	Top 1%	Next 9%	Top 10%	Bottom 90%
Stocks and Mutual Funds+	49.3	40.1	89.4	10.6
Financial Securities	60.6	37.9	98.5	1.5
Trusts	38.9	40.5	79.4	20.6
Business Equity	62.4	30.9	93.3	6.7
Non-home Real Estate	28.3	48.6	76.9	23.1
Stocks, directly or indirectly owned^	38.3	42.9	81.2	18.8
Total Debt	5.4	21.3	26.7	73.4

+ - Directly held (not in retirement accounts)
^ - Owned directly or indirectly (i.e. in retirement accounts)
Source: Recent Trends in Household Wealth in the United States.: Rising Debt and the Middle-Class Squeeze - An Update to 2007; Edward N. Wolff, 2010

The top 10 percent of wealthholders controls over 75 percent of each asset class. Even beyond that, for most asset classes the wealthiest 1 percent typically owns over half of what is controlled by the top 10 percent.

Since these capital assets receive the excess profits produced by our economy, having this unequal distribution of wealth ownership could cause future income disparity problems. The chances of this happening would be reduced by the economic policies advocated in this chapter, because they would reduce the level of excess profits we currently have in the economy and curtail the amount of take-home income received at the highest reaches of the income ladder.

But income disparity isn't the only problem caused by disparate asset ownership, nor is it the only reason to support a reasonable level of estate taxes.

The driving principle behind the existence of the estate tax is one of the same ones upon which this nation was founded: hereditary power (in the form of a monarchy or inherited wealth) breeds corruption and tyranny, while self-determination and democracy produce a more prosperous and fair society. Thomas Paine, the Father of the American Revolution and patron saint of libertarians everywhere, railed against inherited power in his 1776 pamphlet, *Common Sense*, writing, "All hereditary government is in its nature tyranny." He continued, "Hereditary succession . . . is in its nature an absurdity, because it is impossible to make wisdom hereditary." Paine later expanded this critique to inherited economic power in *Agrarian Justice* (1797), where he called for an inheritance tax.[*] While he harbored a great distrust of government and tended to dislike taxes, he understood their necessity, calling taxation the "criterion of public spirit" in his 1782 pamphlet *The Necessity of Taxation*.

But the father of the American Revolution hasn't been the only important figure in American history to favor an estate tax. Andrew Carnegie, one of the wealthiest men of the late 19th and early 20th centuries, realized that it's not a wise thing to leave massive amounts of unearned wealth to children who have not worked for it. In fact, Carnegie testified before Congress in favor of an estate tax. He also donated over 90 percent of his estate before he died, leaving a modest trust fund to his family. In his 1889 tract *The Gospel of Wealth*, Carnegie wrote:

> "Why should men leave great fortunes to their children? If this is done from affection, is it not misguided affection? ... it is no longer questionable that great sums bequeathed often work more for the injury than for the good of the recipients."

That was also the finding of researchers Thomas Stanley and William Danko. In *The Millionaire Next Door* they show that the more inheritance adult children receive, the less they accumulate on their own; heirs of fewer dollars tend to accumulate more. They also

---

[*] In the same pamphlet, Paine also argued in favor of a guaranteed minimum income.

found that receiving a large inheritance was the single most important factor explaining the lack of productivity among adult heirs of the affluent.

Warren Buffett, one of the richest (and wisest) men in our present generation, has been one of the most ardent and outspoken supporters of the estate tax as well. Buffett spent years taking advantage of his investment prowess to amass $44 billion in wealth. And yet he will leave his children less than one-tenth of that amount — not because of taxes, but because he realizes that if talent can't be passed to one's descendants, money shouldn't be either. Buffett equates a repeal of the estate tax to "choosing the 2020 Olympic team by picking the eldest sons of the gold-medal winners in the 2000 Olympics."[34]

Most importantly, inherited wealth undermines our free market system itself. Buffett has argued, "Without the estate tax, you in effect will have an aristocracy of wealth, which means you pass down the ability to command the resources of the nation based on heredity rather than merit."[35] Buffett realizes that economic productivity requires competitive markets, which require a level playing field — something that massive sums of inherited wealth tend to abolish. This echoes the sentiment of Franklin Roosevelt, who claimed:

> "Such inherited economic power is as inconsistent with the ideals of this generation as inherited political power was inconsistent with the ideals of the generation which established our Government ... A tax upon inherited economic power is a tax upon static wealth, not upon that dynamic wealth which makes for the healthy diffusion of economic good."[36]

Even Republican President Theodore Roosevelt was in favor of an inheritance tax, proposing it before Congress in 1906.[37]

Regardless of the arguments for or against an estate tax, the reality is that few people are ever even affected by it. And for those who *might* be affected, there are numerous ways to avoid or reduce the amount of tax owed. To see what I mean, consider some figures from 2003. In that year 2,448,288 Americans died. Of those 2.5 million, only 62,718 (2.6 percent of the deceased) had an estate big enough that they *might* owe some estate taxes. Of that select group, only 30,276 (1.2 percent of the deceased) ended up having to pay any tax at all.[38]

It's obvious that there are important economic and social consequences that favor the existence of an estate tax. But it is also fundamentally unfair to tax the lifetime collection of wealth a person has made, so these competing interests must be balanced. Those who choose to save and create a better life for their heirs should have the right to do so without the fear of having those savings taxed away by the government. However, since that wealth has been gained, at least to some degree, from the exploitation of workers or through imperfections in the economic system, and its disposition threatens the proper functioning of the capitalist economy, it also makes sense that it should be limited to some degree, or some of it should be returned to society.

With these competing interests in mind, we should re-establish a reasonable estate tax for the United States. The estate tax in place before the Economic Growth and Tax Relief Reconciliation Act of 2001 (with an exemption of $675,000 and a maximum rate of 55 percent), seems much too punitive. However, having no estate tax (as was the case in 2010) doesn't make any sense either. Congress recognized that and in 2010 created a new estate tax during their extension of the Bush-era tax cuts. The new law creates an exemption of $5 million and a taxable rate of 35 percent. While the exemption figure represents a reasonable amount — Democrats wanted an exemption of $3.5 million — the tax rate should have been higher, probably in the range of 60 percent.

But given the arguments above surrounding the threats to our democracy from the overabundance of economic power in the hands of a limited few, this is one place where per-heir exemptions would make a lot of sense. Instead of having a flat exemption amount, such as $5 million, we should base the exemption on the number of heirs receiving the estate. For instance, with a per-heir exclusion of a million dollars, a benefactor with four heirs would be allowed a $4 million exemption. (Any amount up to a million dollars given to an heir would be exemptible, but that amount would actually have to be given; the estate could not give an heir $10,000, but then count them as a million dollar exemption.) The remainder of the estate would be taxed using a simple, progressive system much like the income tax. The first $10 million (above the exemption amount, whatever it is for that particular decedent) would be taxed at a rate of 40 percent.

Anything bequeathed above that amount would be taxed at a 70 percent rate. This would encourage a wider disposition of assets and ensure that an extremely large quantity of economic power doesn't fall into the hands of one citizen.

Adjusting the Supply and Demand for Labor

Gradually lifting the minimum wage and implementing income and estate tax systems that make sense will help alleviate our income disparity situation. Yet there remain a few other things that the government can do to reduce income disparity levels that don't lend themselves to specific, quantifiable policy recommendations. Obviously, since we are trying to raise the wages of ordinary workers, anything that reduces the labor supply or increases the labor demand works in our favor. To that end, the United States should take steps to eliminate illegal immigration, support a proper role for labor unions in the workplace, and limit the number of mergers and acquisitions allowed by the Justice Department.

There has been a large debate as to whether illegal immigrants help or hinder our economy. Those seeing a benefit have argued that most illegal immigrants pay taxes, spend money, and contribute to the economy. Opponents have countered that "illegals" actually cost the U.S. economy by taking advantage of benefits, such as a free education and emergency room care, paid for by American citizens. Even if it were true that *all* illegal immigrants pay taxes and *none* of them ever used any public services, there would still be a problem with allowing undocumented workers into the United States.*

The ultimate problem is that every illegal immigrant who obtains work in the United States takes away a job that could be used to employ an American (who would also pay taxes and contribute to the economy). If the country were at full employment, this might not be a problem, but with approximately 15 million citizens out of work, the excess labor coming over the border is only making the matter worse — lowering wages and forcing Americans onto the public dole. In the 20 years between 1980 and 2000, immigration

---

* There are also obvious national security issues with regards to unknown immigrants entering the country. However, for this discussion I will limit the discourse to the economic effects of illegal immigration on native labor.

elevated the supply of male labor by 11 percent in the United States, and caused a 3.2 percent fall in the wage of the average native worker.[39] Of course, these depressive income effects most affected the least skilled and lowest paid Americans, the ones most hurt by income disparity.

According to George W. Bush (the younger), these migrant workers are simply taking jobs that Americans don't want. But the assumption that illegal immigrants only come to the United States to pick fruit and perform agricultural tasks is absurd. Of the 11 million or so illegal immigrants we presently have within our borders, less than one million work on farms for more than half the year[40]. The rest work in construction, landscaping, and a host of other jobs that Americans would love to have. As for the farm jobs that need to be filled each year, if the economy were functioning properly and our nation's income were more equitably distributed, farmers wouldn't have such a hard time making ends meet and there would be less of a need to hire cheap immigrant labor.

There is no reason we can't stop illegal immigration. It even makes sense that this would fall under the "national defense" mantle and could be handled by properly trained military personnel stationed along the border. And although it would be prohibitively expensive to round up and expel all the undocumented immigrants in the United States, doing so would obviously go a long way to creating employment opportunities for America's unemployed.* When illegal aliens are found through other routine means, they need to be deported, and the border needs to be secure enough to ensure that they can't come back. Most of all, the last thing our government should do is give amnesty to those already here or provide a means for them to take advantage of their situation to earn legal citizenship.

In addition to reducing the supply of labor available, we should also strengthen the bargaining power of workers. One of the reasons for the reduced income disparity that we witnessed after WWII was the growth and activism of labor unions. Unions help large numbers of laborers speak with one voice and act in concert in the economy. This concentration of power gives workers more

---

* Note that if the country ever reached the goal of full employment, or even came close, there would be no need (theoretically) for the minimum wage. At full employment, the market – acting more like the theoretical model – would take care of the low wage problem itself.

pricing power when selling their labor and helps offset the concentrated power that companies have when hiring. While raising the minimum wage can help boost income on the lower end of the scale, its effectiveness for raising wages is diminished as we climb the income ladder. Unions can help fill this void by boosting wages for the middle class.

I do not, however, unreservedly endorse all labor union activities or positions. Violence and intimidation by union members and officials should never be condoned or tolerated. Nor should support be given to union demands for concessions that could at some point hinder a company's ability to manage its operation. Employers must have the ability to fire unproductive or problematic employees and reduce staff during economic downturns. At the same time, the government should support union demands for a safe workplace and good pay and benefits. Ultimately, workers need the ability to organize and fight for proper compensation and working conditions, not the ability to hamstring management.

Another important way the government should show its commitment to its citizens is by having the Justice Department strictly enforce antitrust laws, particularly when it comes to mergers and acquisitions. Any time businesses merge, or one business buys another, competition is reduced and the new entity gains pricing power. If the marketplace is going to work properly it must have robust competition from many buyers and sellers. Allowing consolidation of competitors in an industry obviously prohibits a high level of competition.

These business marriages not only hurt consumers, but the new entity usually sheds jobs during the amalgamation, reducing employment. Having fewer employers in a given industry also gives companies more power in wage negotiations, whether with an individual or a labor union. Increasing the size and power of companies through mergers and weakening the bargaining power of workers advances the movement toward higher levels of income disparity.

As if this weren't detrimental enough, too much consolidation also helps create businesses that society comes to view as "too big to fail." This "too big to fail" phenomenon has received considerable attention in the last few years as the government stepped in to rescue numerous firms on the verge of collapse. In the sole instance that the government didn't take action, the failure of

Lehman Brothers, the financial markets were sent reeling and credit markets froze.

But asking whether or not a particular business is "too big to fail" misses the real point. The question we should be asking is, "Who allowed this business to get so big that its failure could bring down the entire economy?" If the failure of a given business could cause a collapse of the economy as a whole, that business is controlling far too much of the market in which it's involved. And it's not as if these "too big to fail" entities got to their enormous size through simple, organic growth. Most of them came to be as big as they are through mergers and acquisitions (M&A).

Consider Citigroup, given a federal bailout of tens of billions of dollars, as well as loan guarantees of over $300 billion. Citigroup was created when Citicorp and Travelers Group merged, but each of those entities was a conglomeration of formerly independent businesses. Citicorp started in 1812 as the City Bank of New York. On its way to becoming Citicorp, it swallowed the International Banking Corporation, the California Federal Bank (previously merged with First Nationwide Mortgage Corp.), the First American Bank of Bryan (Texas), and the Narre Warren-Caroline Springs credit card company, not to mention merging with the First National Bank. The Travelers Group side of the relationship started with Commercial Credit, which bought out Primerica (which had previously purchased insurer AL Williams and stockbroker Smith Barney), and then acquired Travelers Insurance in 1993. That same year, retail brokerage Shearson Lehman was acquired. Four years later, Travelers (as it was then called) merged with Aetna Property and Casualty to become the Travelers Group before buying bond dealer Salomon Brothers. Lost yet? The story gets better. When Citigroup and Travelers Group merged in 1998, banks weren't allowed to own or be merged with insurance underwriters *or* investment banks. But before the deadline for the new entity to divest of the conflicting assets, the regulation prohibiting the conglomeration was repealed with the Gramm-Leach-Bliley Act of 1999 (no doubt with considerable lobbying by Citigroup).

And Citigroup isn't the exception; they're the rule. Look at any firm that took bailout money from the government in the last few years and what you'll find is a conglomeration of what were formerly independent businesses. If the government had prevented most of the mergers and acquisitions that created these behemoths, it would have been easier to allow the poorly managed businesses to fail.

It might be foolish to think that preventing mergers and acquisitions will completely thwart any business from ever becoming too big to fail, or even that all M&A activity should be prevented. But our leaders have an obligation to "promote the *general* welfare," not the welfare of a limited few, and it's apparent that a lot of the business mergers today could cause future harm to our economy in numerous ways. Our government needs to be more rigorous in determining which it will allow and which it will prevent.

As I mentioned earlier, none of these final guidelines fit nicely into specific policy proposals, but all are important nonetheless. Along with the more precise suggestions mentioned above, they help to alleviate the income disparity problem by rectifying domestic imperfections in our economy. But because our economy is increasingly becoming part of an international market, we must address the effects of our trade policies as well.

A New Trade

The simple changes outlined so far should go a long way to rectifying the economic problems in our country and reviving the economy, but they won't be enough. The United States has lost a large number of manufacturing jobs that at one time were the backbone of our economy. In the mid 1960s, manufacturing accounted for 53 percent of the American economy; in 2004 it comprised just 9 percent. This loss of manufacturing employment has been a major contributor to the reduced wages for average workers, the unemployment, the budget deficits, and the weaker dollar we've experienced over the last few decades. If we're going to have a vibrant, diversified economic future, America must regain that manufacturing base.

But that begs the question: what happened to cause our country to lose its production base to begin with?

The answer has two parts. The first is that numerous assembly-line jobs have been lost to automation. Machines have taken over many of the jobs that involved simple, highly repetitive tasks. And while this trend might have been somewhat unavoidable, the replacement of American production jobs by automation has no doubt been affected by our access to cheap supplies of energy (a subject that will be addressed in the next chapter). But the big job killer, and the focus of the present chapter, is America's poorly designed trade policies. Since the beginning of the 21st century, America has run a trade deficit of no less than $350 billion a year; in a few of those years the trade deficit was $700 billion or more. This is hundreds of billions of dollars' worth of products that we buy from foreign workers that we could, for the most part, buy from our fellow countrymen.

For this entire economic package to work, America must address its international trade problems. To do so properly, we will need to objectively examine both sides of this hotly debated topic. On the one side stand "free traders" — those who believe that all international trade is good and that even if the rest of the world refuses, the United States can still benefit from unilaterally reducing our general level of import taxes. On the other side of the argument are those commonly labeled as "protectionists." Protectionists tend to look at all free trade deals and any lowering of import tariffs as a guaranteed loss of American jobs. As we'll see, neither position is completely right. But in order to see what is wrong with our present position, and develop a trade stance that makes sense, we'll have to dig into both sides of the debate and determine what is good, and bad, about international trade.

The Idea of Free Trade

Let's start by looking at the idea of free trade and the potential benefits it can bring to nations that engage in it.

The theory of international free trade, as first developed by David Ricardo,[#] states that all trading countries can benefit from a

---

[#] Some attribute the original theory of comparative advantage, on which free trade theory is based, to Robert Torrens, but this is outside the necessary discussion here.

reduction of import tariffs. According to the theory, the dropping of import tariffs allows goods to flow more freely between trading nations. As such trade ensues, competition naturally begins between numerous companies that make similar products. This competition produces benefits that accrue to the trading nations. To begin with, the competition results in lower prices, as is usually the case with an increase in business competition. This process can also lead to better quality products being offered. Ultimately, according to the original theory, a country ends up specializing in the product for which it has the greatest comparative advantage, and then trading for everything else. This allows citizens from all nations to enjoy the greatest amount of goods and services for the labor they put forth and raises living standards for everyone engaged in trade. The efficiency of production that develops also ensures that all of the world's resources are used as efficiently as possible.

With benefits such as these, it's hard to believe that anyone would reject the idea of free trade. Unfortunately, these benefits haven't been the ultimate result of our forays into international trade. This is mostly due to the fact that — as with free market capitalism — the theorized benefits are based on a number of assumed conditions that don't actually exist in the real world. So let's examine some of these assumptions and the effect they have on the results we've experienced.

The first assumption we should examine is that there is full employment in every country involved in international trade. This assumption is the same as the one involved in the theory of free market capitalism, except that it applies to all countries. Its effect is ultimately the same: it means that wages in all trading countries are higher than we have seen in the real world. Countries like India and China have an abundance of excess labor so their wage rates are extremely low. Even in the United States, the fact that we have people who would like to work but can't find it means that our wage rates are lower than they should be. Full employment would not only drive up wages, it would make it extremely difficult for companies to move production from one country to another. If a company wanted to move 100 production jobs into a country with full employment, it would have to pay a higher wage rate than those companies already functioning in that country. The whole process could start an upward wage spiral.

Along with the assumption of full employment, there is another reason that companies don't move from country to country in the theoretical world: economists simply assume it won't happen. Countries that have higher wages tend to also have more advanced capital equipment. This equipment helps them be more productive and compete with lower-wage countries, so the country with the lowest wages doesn't necessarily have the greatest advantage. Of course, if a company in a high-wage country can move its capital equipment to a lower wage country, it can have an even greater advantage in the international marketplace. But this also negates many of the benefits of free trade, particularly for higher-wage countries. So under the model of free trade theory, capital equipment can't be moved from one country to another.

There is also no government, and hence no taxes, in the theoretical world of free trade. Note that if a country moves from a policy of taxing imports to a policy of free trade, it will lose the government revenue formerly produced by the tariffs. That means it must make that revenue up by taxing something else. This is precisely what happened in the United States in the early 20th century. Up until that time, the government's main source of revenue had been tariffs on imported goods. As the philosophy of free trade gained wider acceptance and tariffs were lowered, the government needed to find new source of revenue. This new source soon became the income tax that so many people decry today. Of course, just assuming the government doesn't exist — thus there are no taxes — easily solves this loss of revenue problem in the theoretical construct.

Finally, the theory assumes that if there are citizens of a country that suffer a loss on account of free trade, the winners will simply compensate them for the loss. This allows free trade advocates to claim that there is an overall net benefit to having a free trade regime; losers get made whole, and the winners are still better off.

While this isn't a complete list of the assumptions that make up the free trade model, it's certainly enough to show that the theoretical world and the real world are very different. These differences can lead to results that are less than favorable in some situations when countries engage in free trade, or reduce their tariff levels. This doesn't mean we should completely reject international trade, but it does mean we need to take a more realistic view of it in order to come up with a trade stance that makes sense.

Welcome to the Real World

Despite the unrealistic assumptions asserted in international free trade theory, it's important to recognize that there *are* potential benefits to be gained from international trade, even in the real world.

Trade with other nations allows Americans to enjoy products that can't be grown here, or aren't grown here efficiently. Bananas and coffee are great examples. Bananas simply don't grow in the United States, so they must be imported. While coffee does grow in Hawaii, it is unlikely that the island chain could produce enough for our domestic market, so buying coffee from countries like Ethiopia or Colombia makes sense. Allowing these products to be grown in the regions of the world best suited for their culture and then trading for them will cause less of an impact on the environment and let countries like the United States focus on the products we grow best. And allowing them to enter the United States tariff free (or with an extremely low tariff) gives us the opportunity to enjoy these goods at the lowest possible cost.

Competition from foreign companies can force domestic companies to offer better products. The Japanese and Koreans, for instance, tend to produce cars that are more fuel-efficient than those produced here in the United States. Allowing those autos to enter the country provides needed competition to keep the United States' auto industry on its toes. The same can be said about the semiconductor industry, the pharmaceutical industry, the computer industry and a host of others.

Having a healthy amount of trade between countries can also help to promote peace. Nations that trade with or are reliant on another country for certain goods are much less likely to go to war with that trading partner.

All of these benefits are possible when a country like the United States trades with any other country on the planet. They can occur despite the fact that the real world doesn't match the theoretical one. But the differences between the world we live in and the theoretical world of economists can create some problems when it comes to trade. Whether or not these problems arise depends on the relative economic and legal development of the countries engaging in trade.

When countries with similar living standards trade freely with one another, competitive advantages must come from improvements in the manufacturing process. Since wages are similar in both countries, companies must gain a competitive edge by improving their inherent skill, efficiency, or innovation. But that is not the case when a wealthier country trades freely with a country with a lower standard of living. Because companies *can* actually move capital equipment from country to country, manufacturing efficiency isn't the only way to lower production costs. Instead of gaining an advantage through production efficiencies, companies simply move production to the country where labor is cheapest. Even if an advanced country has developed capital improvements that make its production more efficient and helps to offset its higher wage rates, companies will simply move that capital to the country with the lower wage rate. If the workers of the United States are the most productive in the world — as politicians love to tell us on the campaign trail — why do all of our industries continue to ship production to countries like China, Mexico and India? It's simple: they're chasing lower wages.

This transfer of jobs lowers wages in the wealthier country, and raises wages in the poorer country. This phenomenon was recognized early on, in 1748, by philosopher David Hume when he wrote:

> "Where one nation has gotten the start of another in trade, it is very difficult for the latter to regain the ground it has lost because of the superior industry and skill of the former ... But these advantages are compensated in some measure, by the low price of labor in every nation which has not had an extensive commerce ... Manufacturers therefore gradually shift their places, leaving those countries and provinces which they have already enriched, and flying to others, whither they are allured by the cheapness of provisions and labor, till they have enriched those also, and are again banished by the same cause ..."[41]

David Ricardo, the father of free trade himself, admits that businesses will move to areas where they can take advantage of lower wages to obtain higher profits. In his work *On the Principles of Political Economy and Taxation*, he writes:

> "If the profits of capital employed in Yorkshire should exceed those of capital employed in London, capital would speedily move from London to Yorkshire, and an equality of profits would be affected."[42]

This is precisely what happens internationally as freer trade policies are more widely instituted. The only reason Ricardo himself didn't theorize this happening internationally is that he assumed capital would be unwilling to move from one country to another (the two locations mentioned in the quotation above are both in England).

And this happens not solely under free trade regimes; it can occur when existing tariffs are very low. Once trade begins between countries that have different standards of living, corporations will begin to analyze whether or not it makes financial sense to close factories in the wealthier country and move the equipment to the country with the cheaper labor. If the difference in wage rates between the two trading countries is large enough to more than make up for the costs of getting goods back into the more advanced nation (transportation and tariffs), the companies will move.

But it's wrong to think that the outsourcing of American jobs is the only problem. Even if there is not enough of a difference in wage rates to entice companies to move from one country to another, job loss in the wealthier country can still occur. Because businesses in the poorer country can pay lower wage rates, they can make products similar to those made in richer countries, but at a lower cost, and then simply gobble up global market share. Consider the example of BYD, a Chinese maker of batteries, cell phones and electric cars. BYD (the letters stand for the initials of the company's Chinese name) initially was founded to compete with batteries being imported from Japan. (Note that Japan has much higher wage rates than China.) Instead of using robotic arms and machines in the assembly process, like the Japanese, BYD decided to hire thousands of low wage migrant workers to do its assembly work. Although this system is less efficient than the more automated plants of the Japanese, BYD was able to undercut Japanese production costs because the price of labor in China is so low.[43] In these cases, free trade doesn't drive production efficiencies; it creates a race to the bottom of the international wage scale.

In fact, Chinese entrepreneurs are known for going to countries like Germany and the United States, buying products that sell well in those countries, taking the products home and breaking them down to see how they were made. They then establish a factory in China making the very same product at a much lower cost and export it to Germany or the United States.[44] The company in the more advanced country has a hard time competing with these imports and either lays off workers or tries to reduce wages.

This highlights another potential trade problem. Some countries haven't developed a legal framework for business transactions as advanced as those in other countries. Trademarks and intellectual property don't enjoy the legal protections in these less developed countries that they do in advanced nations. That's a serious problem for American businesses. Not only must they compete with companies that can pay much lower wages, but even if they develop and patent a new item, a business in a country like China might ignore the patent and start making the item anyway.

The same problem arises when one country has a lack of regulations to protect the environment, consumers, and laborers. The absence of such regulations reduces the cost of producing goods in those countries, giving businesses that operate there an advantage. Some politicians have acknowledged this and called for worker rights and environmental protections to be part of any free trade agreement signed with another nation. But that begs the question, "How can you recognize the fact that the absence of these types of regulations creates an advantage for a trading partner, but not realize that lower wage rates do the same?" Some might suggest that the solution is for the United States to simply do away with the protections we have established for consumers, workers, and the environment. Once again, this creates a race to the bottom for everyone involved.

All of these problems show that the decision about whether or not to engage in free trade with another nation, or what level of tariffs to impose on their products, should be made on a case by case basis, rather than having the same trade regime in place for all. Mutually reducing import tariffs between nations that have similar living standards and legal systems can improve the standard of living for both countries. On the contrary, reducing tariffs on goods from a country where wage rates are lower and laws fail to protect the environment, workers, and legitimate business interests, is a recipe for job loss and wage reductions.

Debunking "Free Trade for Everyone" Arguments

Despite the obvious problems that have been created from America's reliance on free trade and low tariffs, supporters of free trade still insist it will ultimately work. Rather than finding a way to garner its benefits while preventing its damage, free traders seem more interested in either explaining away the problems it causes, or suggesting government support for those it harms.

Supporters of free trade and lower import tariffs claim that the jobs lost by wealthier countries tend to be in lower paid industries, such as textiles, and the jobs gained tend to be in higher paid industries, such as aircraft and machinery manufacturing. There are a number of problems with this argument. First, even if the jobs lost are lower-wage jobs, they are jobs nonetheless. If they were replaced, one for one, with higher paying jobs, that might be an acceptable trade-off. But this is rarely the case, and the increase in the unemployment rate caused by the loss of jobs puts downward pressure on most other wages. It also reduces the tax base and increases government expenditures on things like unemployment and welfare. Second, it assumes that poorer countries will be able to afford high-end products on low-end wages. For trade to work, each country actually has to be able to buy their trading partner's products. Will Indonesian garment workers be able to afford a $30,000 American-made car on their wages of a couple of dollars a day? Will Chinese shoe manufacturers be able to buy an American-made computer that costs $800 on their salary of less than a dollar an hour? The answer to these questions is ultimately "no." Free trade advocates love to tout the fact that China has over one billion consumers and a growing middle class. What they don't point out is that over a third of Chinese citizens live on less than two dollars a day and being "middle class" means you earn the equivalent of roughly $9,000 a year.* In other words, workers in these countries won't be buying a lot of high-end American-made products.

---

* The China State Information Center considers those earning 50,000 yuan, or $6,227, annually to be in the middle class. Other economists cite higher figures, but none that I have seen are higher than the poverty level in the U.S.

This brings up the final fault in this weak excuse for free trade: countries like China aren't simply making T-shirts and shoes; they're making higher-end products like automobiles and electronics. Caterpillar, DuPont, Apple, General Motors and others have all developed production in places like China (and this is definitely only a partial list). These companies realize that consumers in places like India, Mexico, China, and Indonesia won't be able to afford cars, computers and earth-moving equipment made by high-paid American workers. So in order to sell to consumers in those countries, they've moved production to low-wage countries in order to lower their costs. Many of them are not only using low-wage labor to produce for foreign markets, they're shipping the products back to the United States. A Congressional Research Service report in 2007 stated, "In the past decade, the most dramatic increases in U.S. imports from China have been not in labor-intensive sectors but in some advanced technology sectors, such as office and data processing machines, telecommunications and sound equipment, and electrical machinery and appliances."[45] In other words, we're not just losing garment factory jobs to places like China and Mexico; we're losing well-paying manufacturing jobs as well.

Authors like Thomas Friedman have tried to address this issue by suggesting that Americans simply need to bolster their education and job skills so that our country can win those high-paying jobs of the future that require expertise in math, science, or engineering.[46] If we were losing high-end manufacturing jobs solely due to a lack of education, then more education would be the only answer, but that's not the reality of the situation. Friedman's advice is worthwhile if you are an individual strategically planning the rest of your life. Having a good education can definitely help you win a good job and increase your income. But his advice is almost worthless if you're the President and you're trying to ensure that our economy functions properly for the next 50 years. To see what I mean, imagine that we are successful in following Friedman's advice: we steal all the high-paying jobs in computer engineering, bio-engineering and aerospace technology (or whatever the "up-and-coming" industries turn out to be), and leave lower-paying jobs for our foreign competitors — trading our high-tech goods for clothing and shoes (once again, assuming that workers in poor countries could afford these products). Great. Now who grows our food? Is it done domestically, or do we outsource that, as well? If it's grown domestically, how much do the farmers make? Do they live in poverty because they don't have high-tech jobs? (If it's outsourced, what are foreign farmers using to fertilize the food? Is the water they use to grow the plants safe for consumption?)

You see, a properly functioning, robust economy needs a full range of jobs across the entire income spectrum, and if the low end of the wage scale falls too far behind, the economy will simply cease to function properly. So regardless of whether or not we "win the jobs of the future," we still must address our income disparity problem. And having corporate America export jobs — regardless of the industry — in order to take advantage of lower wages elsewhere makes that almost impossible. Rather than endangering our national security* and having the entire American workforce continually going back to school to try to learn the latest high-tech skill — while wages continue to fall because everything that needs to be manufactured gets outsourced — why don't we just admit that free trade with every nation doesn't work and put a simple trade plan in place that offsets the wage and legal advantages that undeveloped nations have?

Free trade proponents also tend to fall back on their claim that trading with poorer nations is not a problem because there are automatic balancing mechanisms in the world economy that counteract any trade deficit our country might accumulate. The first of these is a change in currency values. As Americans import goods from a country like China, we must sell dollars (lowering the value of the dollar) and buy Chinese yuan (raising the value of the yuan).+ As the dollar falls and the yuan rises, American goods become cheaper in China and Chinese goods become more expensive in America, thus boosting our exports to China and reducing Chinese imports to the United States.

The second balancing mechanism is known as "balance of payments accounting." This principle states that a country's purchases of foreign assets and products must equal the sales of its

---

* Which is precisely what would happen if tasks like growing our food were outsourced.

+ Actually the companies who do the importing make these currency transactions, but the effect is the same.

own assets and products to foreign entities. The balance of payments breaks down into two categories: the current account and the capital account. The "current account" includes all trades in goods and services, while the "capital account" includes trade in assets (e.g., stocks and government bonds). Balance of payments accounting theory contends that if we run a trade deficit with China, by buying more of their goods than they buy of ours, China will buy an amount of American assets that equals that deficit.

There are serious problems with both these balancing mechanisms. First, countries like China manipulate their currency, thus eliminating the self-correcting foreign exchange mechanism. Even when countries don't blatantly peg the value of their currency to the dollar, they are constantly expanding and contracting their money supply, which affects the exchange value between the two currencies.# This is compounded by the fact that the sheer volume of foreign exchange transactions that occur in a year's time far outweighs the level that would be necessary to transact foreign trade. In other words, currency markets aren't really free markets that are affected only by the trade of foreign goods and services. Because a lot of speculation occurs in foreign exchange markets, it is highly unlikely that they will produce the balancing mechanism expected. And even if the currency adjustment system did work perfectly, it's obvious that the U.S. dollar would have to fall dramatically in value in order for us to produce a trade balance with the rest of the world. Is that what we want: a weaker currency and a lower standard of living in comparison with other countries on the world stage?

The trend demonstrated by balance of payments accounting — foreigners taking the dollars they earn when we buy their products and investing them in America assets — is less than ideal as well. Free traders love to praise this development, as if value is demonstrated by the fact that foreigners want to invest in America. But as we saw in Chapter 2, the owners of capital receive the excess profits created by businesses. If we run a trade deficit with a foreign country and they reciprocate by buying our assets, then these excess profits go to those foreign investors (or we end up paying a fair portion of our taxes to them because they own so many of our

---

One could easily argue, with the flood of new dollars printed by the Fed in the last few years, that the United States is manipulating its currency as well.

government's bonds). Becoming low-paid workers who create massive profits for foreign capitalists is hardly the way to build a strong and prosperous America.* Yet this is precisely what is happening because of America's unsound trade policies.

Additional problems with the idea of free trade arise because most countries subsidize their production of various products. These subsidies reduce the manufacturing costs of these goods and thus make them more competitive in the world market. This again brings us back to the problem of winners and losers on the world stage being determined by production costs that are unrelated to manufacturing efficiency, instead of being determined by which company has developed the most efficient production process.

All of this gets swept under the rug or explained away by those who support reducing tariffs in every instance. Growth, brought on by the wonders of free trade, easily outweighs any negative effects a country may experience, according to their argument. A typical example of this can be found in Douglas Irwin's book *Free Trade Under Fire*. In it, Irwin cites three examples — South Korea, Chile, and India — of countries that liberalized their trade policies and then saw greatly improved economic growth.[47] Irwin claims that the enhanced growth rates were due to freer trade policies. But he mentions, in passing, that two of the countries — South Korea and India — devalued their currency in the process of enacting the new trade laws. Well, of course these countries are going to experience greater exports and economic growth! As I mentioned a bit ago, when a country's currency falls in value (i.e., devaluation) its exports become cheaper on the world market, thus lifting their demand. Irwin seems to ignore the devaluation effect and he attributes all of the growth to the change in import tariffs. The example countries he cites also didn't have living standards that were high enough to be easily undercut by low-wage competitors, so as long as each country's industrial base was mature enough to handle international competition, one would have expected them to benefit quite nicely from the enacted changes. It's also important to note that in each of these three cases tariffs were reduced, *but not*

---

* Warren Buffett, quoted by the Associated Press in January 2006, made the point, "The U.S trade deficit is a bigger threat to the domestic economy than either the federal budget deficit or consumer debt and could lead to political turmoil ... Right now, the rest of the world owns $3 trillion more of us than we own of them."

eliminated. So if these examples prove that a reduction of tariffs can enhance economic growth in these countries, they also show that countries can experience good growth with modest tariffs in place.

Comparing free trade among nations to the trade carried on between states in the United States is also a tactic used by proponents of free trade. The difference between interstate trade and international trade is that the United States is all under one currency, one set of environmental regulations, and one set of labor laws.* If the entire world operated under the same conditions, there might be a better argument for free trade. But even in the United States we see some microcosms of the problems that would be incurred with completely free international trade. States regularly lose large businesses to other states due to differences in wage levels, rules regarding unions and labor, as well as educational systems. A perfect example of this is RCA, which moved production twice within the United States, before moving it to Mexico, all in search of more submissive, lower-paid labor. Moreover, the federal budget has established a system where tax revenue from more well-to-do states benefits citizens in states that are not as well off. This blurs the distinction between the winners and losers in interstate trade.

This brings us to an ugly point about free trade often ignored by its supporters: the need for international governing bodies and the loss of national sovereignty. In order for international trade to work properly, each participating country must yield to a set of international rules regarding trade — as well as the international body that adjudicates these rules — bringing us one step closer to the "one world government" that so many Americans detest.* The United States has already sacrificed a good deal of its sovereignty simply by being a member of the World Trade Organization (WTO). Laws passed by our democratically elected leaders here in America have already had to be overturned in order to satisfy our commitments to this international trade organization. In fact,

---

* Although there are some differences in labor and environmental laws from state to state, these tend to be minimal.

* It's ironic that conservative pundits often decry America's involvement in the United Nations out of concerns for our national sovereignty, but never rail against our participation in the World Trade Organization, even though our national sovereignty is much more at risk from the WTO than it is from the UN.

international trade regulations have already so permeated the world that countries can't even resuscitate their economies without worrying about breaking international trade laws. During the debate and passage of the economic stimulus bill in the late winter of 2009, much was made about whether the "buy American" clauses in the bill would break our commitments to international trade institutions. Obviously, if our government enacts a spending program in order to pull the economy out of a recession — or in this case, a possible depression — it would be best if the stimulus money was spent within our country on American-made products. But this "discrimination" against foreign companies might violate our commitments to the governing body of international trade. In other words, U.S. taxpayer money may end up going toward pumping up the economies of other countries rather than our own. This type of situation — created by free-trade idealism — is simply ridiculous.

Because it's obvious that some Americans are harmed when we lower tariffs on goods coming from certain countries, and the poor excuses provided by free traders haven't justified the losses, our government has implemented "trade-loss adjustment" programs. The assistance provided by these programs helps to compensate or retrain those who lose jobs due to free trade. The idea is based upon the "winners compensate the losers" assumption in modern-day free trade theory. Free traders have convinced our government that America, as a whole, must be benefiting from free trade, so some money should go to compensating those who have lost their jobs because of it. Of course there's no mechanism in place to iensure that these funds come from the "winners" of free trade. In fact, to the degree that free trade has driven down the wages of American workers, it has helped businesses and business owners and made them the beneficiaries. But tax rates on these groups haven't increased, they've fallen. This would seem to suggest that trade loss assistance programs are more about mollifying those who are harmed by our trade policies, than actually setting up a system that would show whether or not America is seeing a net benefit from them.

A Policy That Works

The issue of international trade is complex and difficult. It's obvious that there are benefits to be had by lowering trade barriers on goods coming from certain nations. It's also apparent that allowing goods into the United States tariff free, or with a very low tariff, from countries with wage rates lower than our own provides an unfair advantage to companies located in those countries. In some instances it even invites the outsourcing of our jobs. Real solutions, given this complex set of costs and benefits, have been elusive.

Complicating the whole situation is the "800 pound gorilla in the room," better known as China. As you probably noticed, most of the examples of international trade problems I gave in this chapter involved the Chinese. While we've traded with China for decades, our trade deficit problem with that country didn't begin to develop until we began bestowing them with Most Favored Nation trade status on an annual basis in the 1990s. The wheels really came off when Congress decided to permanently grant them that status in 2000. Most Favored Nation trade status greatly lowers American tariffs for countries granted the designation, and the results of China receiving this label are proof positive that we should not have the same tariff rates on imports from every country. Not only does China benefit from an abundance of extremely cheap labor, their manipulation of their currency greatly amplifies that advantage. While America has run trade deficits with various countries in the past, and still does in many instances, it has never faced an imbalance like the one we currently have with China. The situation tempts those who believe in the value of tariffs to suggest draconian measures against all countries with poorly paid workforces. Fortunately, that isn't necessary.

What *is* necessary is a realistic assessment of the problems surrounding our foreign trade patterns and an unbiased proposal for how our trade policies should be changed.

The first thing we must do is admit that some of these conditions — created by trading with countries that can produce cheap products, regardless of their efficiency or inefficiency — are enhanced by some of our own problems. The income disparity and wage stagnation that we've been discussing up to this point forces a majority of Americans to look for cheaper models of the products they buy. This, in turn, has been compounded by businesses like Wal-Mart, whose astounding growth and popularity has come from providing ever-cheaper prices, often by importing their products from countries with cheap labor. The economic plan proposed here rectifies this problem by correcting America's income disparity problem and insuring that more workers in the United States earn enough to "buy American."

But fixing our internal problems won't provide a complete solution. We must develop a trade policy that allows the multitude of benefits brought on by free trade, while protecting American workers from the unfair competition presented by low-wage competitors. The best way to do this is with a multi-tiered tariff system, with tariff levels for our trading partners determined by how similar their wage rates, environmental standards and legal systems are to ours in the United States. A system like the following makes the most sense:

1. Goods imported from countries that have a standard of living roughly identical to the United States, or higher, should pay a minimal tariff, say 1 percent. This rate could apply to any country whose average wage is 75 percent or greater of the average American wage and whose laws don't differ materially from our own. Examples of countries in this category would be Qatar, Canada, Norway, Australia, Switzerland, and Ireland (to name a few).
2. Goods coming from countries with an average wage between 50 percent and 75 percent of the U.S. average would be charged a tariff in the 4 percent range. This would help to offset the slight wage advantage these countries have. This category would include countries like Germany, Japan, New Zealand, Spain, Israel, and South Korea (again, only a partial list).
3. Goods imported from countries with wage rates that tend to be 25 to 50 percent of the American average would face import duties of 8 percent. Since these countries have a greater wage advantage than those mentioned above, their goods would face a higher rate. This rate would apply to goods from countries like Turkey, Mexico, Argentina, and Chile (as examples).
4. Products imported from countries with extremely low wages (less than 25 percent of U.S. wages on average) or poorly developed regulatory systems would be charged a tariff of 12 percent. Indonesia, Vietnam, Bolivia, Nicaragua, and India — most of which have a per capita GDP one-tenth or less of

that in the United States — are examples of countries that would fall into this category.
5. Goods imported from China (or any other nation found to be manipulating the trade environment) would face a tariff in the 20 - 25 percent range.
6. Any products that can't be grown or produced in the United States would be imported at the 1 percent tariff rate.
7. All quotas currently in place would be eliminated.

Let's look at the logic behind this type of system, and its expected benefits.

The 1 percent rate, mentioned in points 1 and 6, is essentially free trade. The only difference is that this minimal rate would help pay for the logistics involved in the importation process. Government money must be spent providing port security and inspectors for incoming cargo, which we continually hear are in short supply. Security must also be provided for shipping lanes by the U.S. military and Coast Guard. This 1 percent tariff would to some extent offset these costs. In other words, it would make those who demand the imports pay for the true cost of getting the goods safely into this country. This rate would apply to any product that couldn't be made or grown in this country, or that comes from a trading partner that essentially has no wage or regulatory advantage over the United States. This rate wouldn't inhibit trade, and would therefore allow the benefits of international competition, but would provide the funds needed for the importing operation.

Points 2, 3 and 4 simply impose gradually higher tax rates on goods coming from countries that have progressively lower wage rates, or less developed regulatory systems, than those found in America. The lower a country's wage rates, the greater the advantage that trading partner has over the United States. This rate system would help to offset these advantages. These rates aren't excessive, but their progressivity would be recognition that there is an unmerited danger to American workers from workers in lower-wage countries.

Point 5 simply offsets the damage being caused, not only by China's low wage workforce, but by their currency manipulation as well. If manufacturers wish to take advantage of China's abuse of our international trade relationship, they should have to pay the price when they ship the goods back to the United States.

The seventh point rightly eliminates all import quotas. Quotas don't work in the same manner as tariffs, or produce the same results. Placing a quota on imports raises the price of those goods for American consumers, just as a tariff does. But the benefits derived from the quota accrue to different recipients. Under a tariff regime, the economic benefit accrues to the government and domestic industries. (Preferably tariffs are kept low so that there is no unnecessary benefit to domestic producers, merely a more level playing field between them and foreign competitors.) With a quota, the economic benefit that comes from the restricted supply accrues to domestic suppliers and those foreign producers fortunate enough have their goods included in the quota. Since foreign governments usually get to choose which of their country's companies get to sell in the quota-restricted market, quotas tend to cause political corruption as vendors try to bribe their way into inclusion in the quota. This change would eliminate those negative effects.

If the thought of such an import tax system sounds confusing or complex, simply take a few days and look over the latest "Harmonized Tariff Schedule of the United States" (HTS).[48] The HTS is published on a yearly basis by the U.S. International Trade Commission and outlines tariff rates for all of the imports entering our country. Although completely reviewing it would take a number of days, it only takes a few minutes' scan to realize that there is no logical basis behind its bewildering number of rates.

Even worse than its complexity, our present tariff system sets the import fee based on *product*, not *country*. This has the effect of protecting some products and industries more than others, which flies in the face of free trade theory and the aspects of it that are actually beneficial. Instead of protecting *certain industries* against competition from *all countries*, we should be protecting *all industries* against competition from *certain countries* — countries that have unfair production advantages due to their wage rates, environmental laws, legal rights, and so on. This plan would open all domestic industries to fair competition from countries with policies like our own, while protecting them from ones with different policies.

This system is simple and straightforward, even more so than most free trade agreements. If free trade agreements were really about producing tariff-free trade, they'd only be a few pages long. (How many pages does it take to explain that each signatory nation will allow goods from the other signatories into its country tariff free?) Instead, most "free-trade" agreements are thousands of pages long, indicating that each side has actually negotiated numerous concessions that skirt the creation of what would actually be tariff-free trade. This system shows no favoritism to a particular industry or business, but treats them all the same.

The United States could even make exceptions for countries like Pakistan or Afghanistan, allowing imports from those countries to enter the country at the 1 percent rate. This would stimulate the economies in those countries and provide jobs to young men and women, keeping them from turning to radical religion and hate as an answer to their problems. In some instances, the United States already does this. But the beneficial effect is muted because of our overly generous trade relations with countries like China, India and Indonesia. Why buy textile products from Afghanistan or Pakistan, where supply chains could be disrupted by violence and terrorism, when they can be gotten just as cheaply from stable countries like China, India or Indonesia? This tariff system would actually boost the effect of our trade-centered diplomatic efforts, and perhaps reduce the need for our military presence in certain areas.

Some might argue that this tariff schedule puts countries in the developing world at a disadvantage; that it relegates their economies to a permanent "Third World" status. Not true. The tariff system outlined above allows any product not grown or produced here in the United States to enter at the lowest rate, encouraging poorer countries to develop their own unique products for export. Because it eliminates quotas, it also doesn't prevent products from entering the United States, as the present system does in some instances. Additionally, nothing in this plan discourages American companies from moving production to other countries, *as long as the goods produced there are intended for sale in those same foreign markets.* Ultimately, this relocation of American production is what is happening anyway under the present trade scheme. This plan would simply discourage companies from making those moves simply to improve their profits at the expense of the American worker.

Companies that have benefited from shipping American manufacturing jobs overseas (and then importing the products of that labor) will no doubt fight this type of change with a barrage of lobbyists. Industries and special interest groups that have enjoyed the protection of quotas will most likely fight the changes as well. But America can't base its trade policy on the benefits of the few, but on the basis of the well-being of the many and what is best for America as a whole.

The difficulty in enacting this type of system lies in the fact that in order to do so America would have to pull out of the World Trade Organization. Advocates of worldwide free trade will no doubt howl at this suggestion, but it will be a necessary step in instituting a trade policy that makes sense. The rules of the WTO require members to give each of their "Most Favored Nation" trading partners the same terms of trade (unless they have worked out specific bilateral or regional trade agreements). This doesn't make sense for a country like the United States, which has more to gain from trade with countries like Switzerland or the Netherlands than we do from trade with countries like China or Mexico. While withdrawing from the WTO might sound radical, the U.S. government has pulled out of multi-national treaties in the last decade for much less noble reasons than saving American jobs, so the precedent certainly exists.

Finally, opponents to this type of change in our nation's tariff schedule might argue that certain nations would view this policy as the first volley of a trade war and institute their own tariff increases in retaliation. Given this plan's elimination of quotas and simplification of our overall tariff system, I would view this as unlikely in most instances. If it did occur, however, the countries most likely to engage in a trade war would be those facing the 12 percent tariff rate or China, which would face a tariff rate of 20 percent or more. Given the fact that over 50 percent of our trade deficit typically comes from trade with countries in this category, a trade war with them might actually force Americans to buy from domestic suppliers and thereby end up helping the U.S. economy.

A Final Note on Trade

Belief in the overstated benefits of universal free trade seems to have overwhelmed our nation's capital. Even President Obama, who questioned the value of free-trade agreements during his campaign for the Democratic nomination, now seems to believe that the way to revive the American economy is by promoting stronger growth of American exports. Any objective observer would have to admit that, yes, increasing exports could help the U.S. economy. But with exports typically representing roughly 10 percent of our GDP in any given year, any rational American would think that finding an improvement for the other 90 percent of our economy would prove much more effective.* But because no one in Washington seems to have an answer for our lagging domestic demand, they all seem to look overseas.

What Washington has failed to realize is that domestic demand has been squelched, not because Americans are tired of shopping, but because an extreme level of income disparity has simply rendered too many of them unable to afford many of the products they would like to buy (without pulling out a credit card or cashing out the equity in their homes). Part of the wage stagnation many Americans face is due, at least in part, to our loss of good manufacturing jobs and our reliance on imports. Our belief in free trade has caused us to implement trade policies that emphasize cheaper production and lower prices at any cost, even if that cost is the loss of American jobs. Expecting American consumers to somehow find the money to purchase goods without also providing them a means to earn a living is a fool's hope. We simply can't import our way to prosperity.

And unless something changes, we won't be able to export our way there, either. In the push for universal free trade, we've failed to ask the most basic of questions: "If Americans are too poor to afford American-made products, how will citizens of *poorer* countries supply a sustainable level of demand for them?" Because this wasn't thought through, our trade relationship with many countries has failed us. Our present tariff system has created a situation where our state and national leaders feel the need to travel the world, begging other countries to buy our products and lower their rates of savings. It's nonsense. Other countries should be allowed to save as much as they like and buy products from wherever they wish. Insisting that they do differently — simply to justify our failed belief in universal free trade — is silly.

If the United States is to regain its industrial base and return to an economic position where it can create jobs for its people,

---

* It's interesting to hear free traders claim that the Smoot Hawley Act of 1930, which raised import tariffs and caused a trade war, brought on the Great Depression. At the time, exports constituted less than 5 percent of the American economy.

instead of simply exporting the ones we already have, it must address its trade deficit and its reliance on the overstated benefits of free trade. We must establish a logical, consistent trade policy that factors in not only the benefits that stem from international trade, but the reality that the world we live in doesn't match the theoretical world of the free trade model. Americans shouldn't be against trade. We should welcome it. But we should welcome it on terms that are fair to the American worker.

Energy Independence and Environmental Sustainability

No in-depth discussion about the long-term health of our economy would be complete without addressing the natural resources and energy needed to keep our economy moving. America has been blessed with a vast array of natural resources — oil, coal, natural gas, gold, silver, vast forests of wood and plains of good soil, gypsum (think drywall), zinc (galvanized products), iron ore (steel), cobalt (used in super alloys), copper (wiring, medical equipment), lead (batteries, bullets), molybdenum (auto parts), and sulfur (petroleum refining), among others. Those resources have helped to diversify and strengthen our economy. Unfortunately, most of them are also nonrenewable. And just as their existence strengthens our economy, their depletion will ultimately threaten it. It doesn't take a doctorate in economics to realize that if your economy relies on a finite set of resources, and those resources run out, keeping the economy moving will become a much more difficult task.

Of course there's a solution to the problem of depleted resource reserves — prices. As a limited resource becomes scarcer, its price rises, sending a necessary signal about the remaining volume of its existing reserves. These higher prices encourage efficiency and a wiser use of the remaining, limited resource. But there are problems with the pricing mechanism.

The first is that the price increases we see usually occur in large, volatile increments as society suddenly realizes how depleted the particular resource is. We witnessed this in 2007 and 2008, as books like *The End of Oil*[49] and *Twilight in the Desert*[50] discussed the declining reserves of petroleum in the world and the price of oil shot from $50 a barrel to over $137 in just 18 months, a 174 percent increase.

These explosive price increases stem from another imperfection with the free market of the real world. If you remember back in Chapter 2, we talked about two of the assumptions that underlie the free market theory, but left the third — the assumption that all market participants have perfect knowledge — to be discussed in this chapter. Obviously, if all market participants — buyers and sellers — had perfect knowledge, there would be greater awareness of how much of a particular natural resource was left and how long that amount would last.* Under that condition price increases would be more gradual. Because the vast majority of people have no idea how much of each limited resource is left, price increases tend to be smaller than they should be, until society suddenly realizes that a given resource is nearing depletion and then the price skyrockets in response.

The second problem with the pricing mechanism is what economists call "externalities." Every time a resource is pulled from the ground and used, whether directly or indirectly, costs are incurred. We see some of these costs when we pay for the products we buy: the price of extracting the materials from the earth, the price of converting those materials into the product itself, and the price of transporting that good to the local market where we buy it. All of these costs typically get reflected in the price of the good. But other

---

* There are other benefits to having all market participants be in possession of perfect knowledge, but this discussion will be limited to the one necessary for this topic.

costs aren't as easily seen because they are not reflected in the price. Economists call these costs externalities because the cost is external to those involved in the transaction.[+] These externalities can take the form of ill effects on our health, as byproducts of these materials pollute our air, water or soil. They can be a disposal cost, in the form of higher local or state taxes that go to pay for a landfill. Or the cost can simply be an aesthetic cost, if the resource ends up as a piece of trash along the roadside, or if the land from which the resource was extracted isn't restored properly. A perfect example of externalities would be all of the costs associated with the devastation to America's Gulf Coast region from the Deepwater Horizon oil spill in 2010.

If the external cost of using a resource isn't factored in, the price we see in the market will be lower than it should be and society will demand more of that resource or product than it should. This is precisely what is happening today: we're using our country's — and indeed, the world's — resources at a faster rate than we should, because the true cost of materials and resources aren't being factored into the price. But don't be fooled; *we are paying these costs* in some way or another. Regrettably, the way we pay for them is usually more expensive and less efficient than it would be if there were simply higher prices on all of the goods we buy.

Some examples will drive home my point. Approximately 30 percent of Superfund cleanup activities are paid for by the U.S. taxpayer, rather than the individuals who bought the products or services that were associated with the pollution or the company that created them. This equates to approximately $1.3 billion a year.[51] Roughly 60 percent of our oil is imported, and much of that comes from places around the globe that are politically unstable (like the Middle East) or have a somewhat adversarial relationship with the United States. Helping to stabilize these regions and protect the shipping lanes through which the oil is delivered has cost the United States trillions of dollars (not to mention the blood of numerous sons and daughters). The funding for these activities comes from the general fund — mostly income taxes — so it doesn't show up in the

---

[+] To be clear, there are both positive and negative externalities. When society experiences *benefits* that are not reflected in the transaction price, this is a *positive* externality. When there is a *cost* to society that is not reflected in the transaction price, this is a *negative* externality.

price we pay for a gallon of gasoline. And a certain portion of the money we do spend on oil from the Middle East ends up in the hands of terrorists who wish to harm Americans, so additional tax funds must be spent on the activities of the Homeland Security Department, which isn't factored in either.

Consider the externalities related to the purchase of electricity that has been made by burning coal. In a report requested by Congress in the Energy Policy Act of 2005, the National Academy of Sciences found that the externalities resulting from emissions of coal-fired power plants amounted to $62 billion in 2005.[52] That's $62 billion of costs that weren't factored into the price of electricity that year. What's even more shocking is that the report didn't consider externalities related to the mining of the coal *or* any climate change effects, so the $62 billion is only a partial figure.

While these external costs are paid for by society in general, they should be paid for by those demanding the good or service whose creation caused the externality. Since these costs are not factored into the price of the good, though, the market becomes distorted and no longer works properly to determine the appropriate level of demand. If they were factored in, prices for these goods and services would be higher, demand would fall, and we would use the associated resources more efficiently.

As we'll see, addressing this issue of externalities is the real key to solving our energy problems and establishing an economy that uses our resources in a more sustainable manner. But before we start discussing possible solutions, and whether they work effectively or not, there are a couple of points that should be made.

Although the long-term depletion of any finite resource poses a problem for our country, it is important to recognize that there is typically a difference between "energy" resources and "non-energy" resources. Most of the energy sources we presently use cannot be recycled; once we use them, they're gone for good. Others, like the materials that go into making paper, cardboard, glass, rubber, and various metals, can be recycled or reused in some way; we can thereby avoid the massive depletion or exhaustion of these natural resources. Therefore, the problem surrounding the use of our energy assets is more immediate than that of using our nonenergy assets. For that reason, the bulk of this chapter will focus on solving our country's energy problem. It should be noted, however, that the approach to solving both problems is essentially the same.

The second point I should make is that the topics of energy use and the environment are closely tied to one another. I've avoided — and will continue to avoid — digging too far into a discussion of the environment because it is such a political lightning rod for controversy and disagreement. As soon as someone mentions energy and the environment in the same discussion the topic of climate change comes up and the discussion quickly devolves into an argument over whether or not global warming exists. I wish to avoid that circus and strictly focus on meeting the challenges created by limited resources. However, any degradation of the environment *will*, to some degree, affect the health of humans. This, in turn, affects our economy, our healthcare spending, the federal budget, and so on. Completely avoiding the topic of the environment in any in-depth, objective discussion of the effects of the nation's energy strategy is not only foolish, but also problematic. Moreover, the Democrats' approach to energy legislation invariably tries to limit environmental damage, so a discussion of the pros and cons of those methods would naturally entail some discussion of the environment as well. Ultimately, there are plenty of "nonenvironmental" reasons to develop a better approach to the use of our natural resources, without distracting some readers with the controversial subject of environmental destruction. As we'll see, the best approach will address these "nonenvironmental" factors and the environmental ones as well.

The Wrong Solutions

Before we get into an appropriate solution for our energy problems, let's examine what the Republicans and Democrats have proposed, and why those solutions aren't really adequate.

The Republicans' approach to our energy problems, particularly our need for oil, stems from the viewpoint that the problem is one of supply and source, rather than one of demand. Whether the trouble comes in the form of extremely high prices, or overdependence on foreign supplies, the answer is to promote more drilling and mining at home. In other words (or in the Republicans' words) if we produced more energy domestically, we'd have to import less, the price would drop, everyone would be happy, and the problem would be solved.

The lack of wisdom and foresight in this "Drill Baby, Drill" approach is daunting. It's akin to a drug addict claiming that his problem revolves around finding the cheapest, most reliable dealer, instead of realizing that the real issue is his addiction.

Let's start by looking at the short term. Creating more energy, or increasing domestic supplies, can't provide us a near-term solution, because it takes a long time to bring additional supplies of energy on line. If we're talking about oil, allowing drilling in new regions, whether in the Alaskan National Wildlife Refuge or off our coastal shores, won't produce usable supplies of energy for approximately 10 years.[53] Natural gas developments face similar time constraints. Additional electricity supplies don't come quickly either. Building dams, coal-fired or nuclear power plants all take time. So if we're looking for short-term solutions, additional production isn't a good one.

To the degree that the resource isn't renewable, creating additional energy from increased resource supplies isn't a good *long-term* solution either. Eventually, the finite resource — no matter what the source — is going to run out. Drilling and digging for additional supplies simply keep us addicted to those limited resources, rather than producing sustainable alternatives. Some have argued, at least in the case of our oil usage, that we could use our domestic supplies while we develop renewable alternatives. While this isn't a bad suggestion, the problem with this strategy is twofold. The first is that these domestic supplies won't be available anytime soon, as was just noted. The second is that bringing additional domestic supplies online exerts downward pressure on the price of energy. Low prices discourage efficiency and the development of more sustainable alternatives, the exact opposite of what we ultimately need. In order to move the country from fossil fuels to renewable sources of energy, *which must eventually happen*, the price of fossil fuels has to increase. (We could encourage a move from fossil fuels to renewables by reducing the price of renewables with government subsidies or tax breaks. But this increases the federal budget — something we need to avoid.[*])

---

[*] It's unfortunate, but under the Energy Policy Act of 2005, the oil industry was given subsidies to encourage drilling more, so our tax dollars have actually been going toward continuing our addiction.

Relying on more drilling and mining to solve our energy problems also does nothing to fix the problem of externalities. Our country spends billions of dollars every year counteracting the negative effects of the way we use our natural resources, and none of these expenditures get reflected in the prices we pay for using these assets. Simply pulling more materials from the earth doesn't solve this problem. In fact, it makes it worse.

Maybe most importantly, not only are drilling and digging not good short or long-term solutions, they're more expensive than a lot of alternatives. In fact, it doesn't matter what your preferred energy production method is — hydro, nuclear, solar, drilling for more oil or natural gas, etc. — it's almost always cheaper to conserve a unit of power than it is to produce it. This is true even *before* externalities are taken into account. Of course, conserving energy isn't a huge revenue booster for corporations that produce energy, so there aren't a lot of lobbyists in Washington, D.C., fighting for it.

The bottom line regarding the Republican approach to our energy problem is this: our country's natural resources — oil, coal, ores, timber, natural gas, etc. — represent a strategic and financial store of wealth for our nation. As such, they should be used in a way that truly strengthens the country, rather than inefficiently wasted because it suits our lazy, entitled lifestyle. The resources that this country has been blessed with give us the opportunity to live without dependence on other nations, but only if we conserve them, and use them efficiently. The Republican approach does nothing to drive us in that direction.

But for the sake of argument, let's say we did try to solve the problem by opening up all federal lands to more drilling and mining for the resources they hold. Wouldn't it make sense to gain not only from the benefit of the resource's utility (i.e., the energy it provides), but from its sale on the open market as well?

Not long ago, the Shtokman natural gas field was one of the largest untapped natural gas resources in the world. The Shtokman lies in Russia's sector of the Barents Sea, meaning it belongs to the people of Russia. Russia recently decided to develop the field, and to do so, it chose two private companies, Total and Statoil-Hydro, to assist its nationally-owned gas company, Gazprom. The private companies are helping in the development of the field, but they won't be given a portion of the reserves; they have essentially signed a service contract.[54] Thus the Russian people gain the majority of the benefit of the natural gas taken from their land.

We could do the same with oil produced from the Alaskan National Wildlife Refuge (ANWR). Presently, any time public lands (like ANWR) with abundant natural resources are opened up to drilling or mining, the companies involved pay a small royalty to the government for the privilege of taking the assets, which belong to the American people. We don't have to become communists to see the same benefits as the Russians do, we just have to take the same approach to oil and gas companies. Instead of just giving them the oil and gas that belongs to all Americans (and then letting them sell it back to us), we could simply pay them a service fee for drilling it for us. The government could then sell the oil and use the proceeds for paying off the national debt or reducing taxes. A similar approach was actually taken in Alaska under the governorship of Sarah Palin. Under Palin's ACES (Alaska's Clear and Equitable Share) plan,[55] taxes were raised on oil company profits. Not only was the base rate raised, but the bill imposed a graduated tax scale, so as the price of oil exceeded $52 a barrel, Alaska's share of the profits increased. This ensured that Alaskans benefited from the sale of their natural resources in an amount reflective of the resource's market price.

Unfortunately, nobody in the "Drill baby, drill" crowd has suggested anything like this. They simply propose that we open up public lands and let Exxon, Conoco, or BP have the oil in hopes that this extra supply might bring down the price per gallon of gasoline by a few cents. Their slogan might as well be "Rip us off baby, rip us off!" You don't have to be treehugger to realize how ridiculous this is. You just have to realize that "public lands" belong to all of us.

The proposition that our energy problem — particularly our need for oil — can be fixed by increased domestic supply is pushed by those who benefit from our present reliance on these finite resources. Companies with profits based on mining and drilling will be big winners if the United States remains addicted to fossil fuels and we expand the areas that we let these companies exploit. Dig deep enough into the campaign funding of any politician pushing the "Drill baby, drill" philosophy — or the endowment of any think tank doing likewise — and you'll eventually find someone with deep pockets and a highly vested interest in the drilling or mining industry.

* * *

Most of the proposals that come from the Democratic "side of the aisle" aren't much better than what we get from Republicans. There are a couple of exceptions, and I'll discuss those in a bit. For now let's look at the typical Democratic fare.

Like Republicans, Democrats readily acknowledge that our dependence on other countries for a large part of our oil supply creates problems that must be addressed. Unlike Republicans, Democrats also recognize that there are environmental problems caused by our over-reliance on all forms of fossil fuels for our energy needs. Therefore, most of their methods attempt to address both problems at the same time. The result is usually a "big government" proposal focusing on what forms of energy we use and how efficiently we use them. These proposals take one of two forms.

The first type of proposal involves additional government spending aimed at promoting the use of sustainable forms of energy. These expenditures include low interest loans and loan guarantees for alternative energy projects, grants for research and development, and tax breaks for making energy efficiency improvements and buying energy efficient products. Some liberal politicians and pundits have even suggested a "Manhattan Project" (similar to the one that brought us "the bomb") or an "Apollo Project" (similar to the one that got us to the moon) to develop the next generation of alternative energy.

Although these government spending plans rightly target creating alternative energy sources and making our energy use as efficient as possible, they have a number of drawbacks. The first is that they cost money. If the government had a balanced budget, or was running a surplus, this might not be a big problem. But with the government running a yearly deficit of over a trillion dollars, the last thing we need are solutions that are "budget negative." The second problem is that these programs can distort the decision-making process of market participants, in some cases actually delaying alternative energy development. If the government is considering providing some form of capital for renewable energy companies, those companies might reject private capital, which would most likely require a higher return on investment. Although passing the legislation to provide the government funds could take a substantial amount of time, alternative energy companies that are aware of the possibility of government funds may delay their projects while waiting for the cheaper capital from Uncle Sam. Or maybe someone is developing a very promising technology that would greatly enhance energy efficiency, but can't get government funding because her work is considered too experimental. Instead of continuing in her line of work, she switches her focus to a technology that is less promising, but that guarantees her a chance at financial assistance from the government. This brings up an additional problem: any government assistance is naturally going to favor certain industries, technologies, and energy production methods over others. This puts the government in the very problematic position of having to pick what the best future technologies will be for producing and conserving energy, and who will be producing them. It also creates a situation that breeds corruption and invites an intense lobbying effort on Capitol Hill. With continual changes and developments in technology, what looks like the right choice today might end up being the worst choice a year from now. Allowing these decisions to be made by market forces, rather than the government, would provide more flexibility and a wider range of potential solutions to our energy problems.

    The second tactic usually employed by Democrats is to introduce regulation governing the use of energy, or limiting the emissions caused by the creation of it. The CAFE (Corporate Average Fuel Efficiency) standard, developed in the 1970s to regulate the fuel efficiency of cars, is a perfect example. This standard has the dual purpose of limiting carbon dioxide (CO_2) emissions and reducing gasoline consumption by increasing the fuel efficiency of automobiles. While both of these are worthwhile goals, this type of legislation provides a solution that is less than ideal. The new mandates require a bevy of regulators to enforce the legislation. This additional bureaucracy costs money, once again burdening the federal budget. Bureaucrats are also needed to monitor progress and suggest updates to the government directives on a regular basis. Because these adjustments to the regulation tend to be arbitrary, they promote increases in lobbying activity. The necessary adjustments can also be delayed as politicians spar over how much of a modification to make. To put it simply, these regulations tend to increase the size, cost, and corruption of our government.

Democrats' newest weapon in the battle to limit emissions and drive efficiency, however, is an attempt to avoid the pitfalls of previous tactics. The cap-and-trade system has been hailed by Democrats as the way to fix the energy/pollution/global warming problem through the use of market forces. The cap-and-trade system works by creating a secondary market for carbon emission credits. In other words, the government puts a limit on the total amount of emissions the economy can produce. It then allocates carbon credits, used to offset a company's pollution, to those who are polluting.[#] The credits would then trade in a market, much like the stock market. Companies that develop cleaner ways of producing their good or service, and end up polluting less, could then sell their credits in the market. Environmental groups or concerned citizens could also enter the market and buy credits, thus driving up the price and making pollution avoidance more valuable. Hence a "market" would be determining the value of pollution and pollution abatement. The whole system sounds great ... until you consider the downsides.

"Once a cap is in place, it is very difficult to adjust," says Thomas Crocker,[56] one of the economists who developed the cap and trade model. Any calculation of cost externalities is going to be imprecise. But that's precisely what the emissions cap will be based on. The value of the credits will then rise or fall depending on the market's determination of what the credit is worth. But if the government comes back later and determines they set the cap too low, or too high, and tries to adjust it, they're going to affect the value of the credits and wreak havoc on the market.

Of course, havoc in secondary markets is nothing new. We've seen time and again what traders and speculators have done with the stock market, the currency markets, the credit derivatives market, the junk bond market, ad infinitum. Is this the type of system we want controlling our energy use and pollution levels? David Montgomery, who advanced the cap-and-trade model in the '70s by establishing a mathematical formula for the theory, compounds this

---

[#] The government could also sell these carbon credits when the system is put into place in order to raise money.

point, "You get huge swings in carbon prices with a cap, which creates more volatility and uncertainty for business."[57] The last thing businesses need is uncertainty in their energy costs. Yet given all the speculation that occurs in our present secondary markets, that's precisely what we would get under a system of cap-and-trade. While it sounds good in theory, cap-and-trade is just another casino game that will help Wall Street traders get rich, while everyday Americans pick up the tab and deal with the mayhem.

To hear politicians in Washington discuss it, though, this doesn't seem to be a problem. The problem they tend to cite is the fact that a cap-and-trade system will cause an increase in energy costs to consumers. Of course, if we desire to see greater energy efficiency — as well as energy prices that reflect their true costs — price increases are precisely what we need. How do politicians expect the market to work without an adjustment in the very signal — *prices* — that markets use to direct change? How do they expect the market to produce energy efficiency solutions if those solutions aren't demanded by individuals and businesses trying to avoid higher energy costs? It makes you wonder how much these "proponents of free markets" really know about how markets work.

While the cap-and-trade scheme is an attempt to employ the power of the market, it is a system that is more complex than necessary and presents drawbacks that can otherwise be avoided with better market-based solutions.

The True Solution

As mentioned earlier, the real solution to solving our energy problem, in the short term *and* the long term, is to make sure the market price for any type of energy reflects, at least in some way, the externality costs created by producing that energy. If the true cost of each energy source were reflected in the price consumers pay, the market would determine not only the appropriate amount of energy truly needed, but also the right mix of sources for providing that energy.

The problem is that determining the true amount of externality costs associated with a given energy production technique is always going to be imprecise. While we might be able to pinpoint how much of the defense budget or the transportation budget is necessitated by our reliance on foreign oil, determining that same amount for the homeland security budget would be much more difficult. How much do we spend to combat asthma, bronchitis, elevated rates of infant mortality, and cardiovascular problems, all problems linked to the burning of fossil fuels? What health problems and disabilities stem from having sulphuric acid, mercury, arsenic, and lead in our water, all of which are byproducts of mining operations, and how much do we pay to treat these ailments? Even if we got our energy from nuclear power (a non-fossil fuel source), what would it cost to store the waste and how much would we spend cleaning up a potential leak? We've seen how much damage one leaking oil well in the Gulf of Mexico could do; imagine having nuclear waste in the water supply of Los Angeles or Las Vegas. And how much are we spending on health care because so many Americans are overweight (in part because they no longer use their own energy to perform basic daily tasks)? Whether it's driving to work instead of cycling, shoveling snow and raking leaves instead of blowing them, or taking the escalator instead of the stairs, we Americans have become reliant on our energy-consuming conveniences. I could go on and on, but you get the point: quantifying these costs and determining exactly how much should be attributed to each type of energy source is an impossible task.

But it should also be obvious that the price we presently pay for our energy is grossly understated because we haven't even begun to factor in these externalities. So while we may not be able to determine the exact price we should be paying for each energy source, it's clear that the prices we're presently paying aren't enough. Therefore, increasing the price of energy by any amount, even if imprecise, is a step in the right direction.

The best way to achieve this increase in prices is by taxing the resources that are used to produce the energy. This could be done in a couple of ways, but the best would be through a BTU tax. This is the same type of tax that was proposed by the Clinton administration in 1993. And while Clinton's BTU tax was not specifically introduced to address our energy problems, it and the various versions of a carbon tax recently proposed in the House of Representatives[*] are really the only decent proposals to come from

---

[*] The "Save Our Climate Act" introduced by Rep. Fortney Stark (D-CA) and the "America's Energy Security Trust Fund Act of 2009" introduced by John Larson (D-CT) are both bills that would place a tax on the carbon content of various energy sources.

the Democratic or Republican side in regards to solving our nation's long-term energy situation. Since the Clinton Administration's BTU tax is a worthwhile idea, but didn't become law, it is worth examining why it failed and why it is better than a straight carbon tax.

The purpose of Clinton's BTU tax was to raise revenue as part of his overall deficit reduction program. Although intended to increase revenue, the tax proposal also included the added benefits of improving the energy efficiency of the nation and reducing the amount of environmental damage caused by burning fossil fuels — both additional goals of the Clinton administration. Not surprisingly, the bill met stiff opposition from the energy industry. In addition, Republicans derided the bill as another tax on Americans, one that had the additional disadvantage of being regressive in nature.[#] Ultimately, the proposal was dropped in favor of an almost 14 cent increase in the federal gas tax.

The reason a BTU tax was chosen over a straight carbon tax was to ensure that the tax burden was distributed more equally across various regions of the country. A straight carbon tax wouldn't have included energy created from nuclear and hydroelectric sources. Since a state like Idaho gets a large portion of its energy from hydroelectric dams, it would have had an unfair advantage over a state like West Virginia, which gets most of its electricity from coal. While hydroelectric and nuclear sources are probably preferable to electricity produced from coal and natural gas, they still cause significant externalities that should be accounted for by a tax. Including them in the BTU proposal not only solved this, but made the bill more politically viable.

The 1993 BTU-tax proposal would have levied a tax rate of 25.7 cents per million BTUs on natural gas, coal, liquefied petroleum gases, nuclear-generated electricity, hydroelectricity, and imported electricity. An additional tax of 34.2 cents per million BTUs would have been imposed on refined petroleum products, for a total of 59.9 cents. The tax would have been phased in over three years and then adjusted for inflation after that. While the tax I

[#] Regressive taxes are ones in which poorer individuals are likely to pay as much as wealthier taxpayers. This type of tax typically doesn't bother Republicans, but when you're trying to defeat a proposal, any excuse will do.

propose here is similar to Clinton's BTU tax, there are a number of important differences.

The first surrounds the tax rates imposed on various energy sources. All sources (coal, natural gas, hydro, nuclear, petroleum, etc.) would be taxed at a base rate of 51.5 cents per million BTUs (MMBTU) in the first year. Refined petroleum products would have an additional 68.5 cent per MMBTU tax, for a total of $1.20 per MMBTU. These same amounts would be added to the tax each subsequent year, so that the tax would double in year two, be triple the base rate in year three, and so on, for ten years. By the tenth year, the rate on all energy sources would be $5.15 per MMBTU, with the exception of refined petroleum products, which would have a rate of $12 per MMBTU.

The second difference involves the exemption of hydrocarbon "feedstocks" used in the production of petrochemical products. Under the Clinton proposal, these would have been exempted from the tax. Under my proposal, they are included. These hydrocarbons, found in natural gas and petroleum, are sequestered during the manufacturing process — they don't involve energy production or carbon dioxide emissions — but are still associated with various forms of water and soil pollution. Our use of these petrochemicals in soaps, detergents, pesticides, and fertilizers is unnecessary, and including them in the tax would provide an incentive to move toward ingredients that cause fewer health and environmental problems.

Like the Clinton BTU tax, any electricity produced from completely renewable sources like solar, wind, biomass or geothermal would be exempted from the tax. This would create a price structure that would allow clean energy sources to better compete in the marketplace, without the assistance of government support in the "budget negative" form of subsidies or tax credits. Not taxing these energy sources also makes economic sense, because they don't cause the externalities (environmental, health, and security problems) that the other energy sources do.

Clinton's economic package, that included the BTU tax, also contained increases in government assistance for low income Americans — increases in the earned income tax credit and food stamp program, and creation of various home energy assistance programs — that would have been paid for by part of the BTU tax revenue. The carbon tax presently proposed by Congressman Larson

(D-CT) also has "offsets" that mean poorer Americans will see no real cost from the tax. My proposal has no such welfare. Instead of providing government assistance to poorer individuals to offset the tax, this plan ensures that working Americans make enough to afford it. Even for those earning the minimum wage, the increases proposed in Chapter 4 are enough to cover this BTU tax, so even the poorest of working Americans will not see a reduction in their living standard.

Finally, unlike Clinton's BTU tax plan, this proposal uses none of the revenue from the tax to create subsidies for renewable energy sources or energy efficiency technologies. The levels proposed by the Clinton plan weren't high enough to drive decision makers in the market to choose cleaner fuels and use energy more efficiently. This plan corrects that by increasing the tax level, so that government subsidies aren't necessary to achieve the same result. Instead of relying on government programs, this system puts complete trust in a market where prices more accurately reflect true costs.

The Effect on Americans

Since few people in the United States are familiar with the BTU measurement, it makes sense to translate these figures into terms most Americans can understand. In the first year, the tax would translate into an average of $6.50 of tax per barrel of oil, or roughly 15 cents per gallon of gasoline. It would add anywhere from $7.20 to $12.90 to a short ton of coal, depending on the type of coal. Since that much coal produces about 1,819 kilowatt-hours (kWh) of electricity (once again, depending on the type of coal and the type of plant burning it to produce electricity), rates on coal-produced electricity would increase by less than a penny per kWh. For natural gas, the tax in the first year would be in the ballpark of 53 cents per million cubic feet. The average household doesn't use a million cubic feet of natural gas in a year,[58] so the direct usage cost would be negligible. For electricity that is produced by burning natural gas, the price increase would also be negligible. The additional cost to electricity produced from hydro or nuclear sources would fall somewhere in between the additional costs caused by using coal and natural gas — roughly half a penny per kWh. There may be changes to the underlying price of certain energy sources as they fall out of favor, or see increases in demand, but these changes would have a minimal effect on the price seen by end consumers.

Obviously you can take the figures above, multiply them by the amount of each energy source that you use and get a rough determination of how the tax would affect your wallet, but let's consider some likely examples. Most electric utilities get their energy from a number of different sources, so the tax per kilowatt-hour for electricity will most likely be less than 7/10ths of a cent. But assuming a utility produced all of its electricity by burning coal (the energy source with the greatest effective tax), and passed all of that tax on to the end consumer, the average household (which uses 11,040 kWh per year) would pay $68.45 the first year. In the tenth year, the tax would amount to $684.48 — assuming, once again, that all the electricity came from coal, all the tax was passed through to the end consumer, and the household hadn't changed any behaviors to increase the efficiency of their energy usage (all three highly unlikely assumptions). Obviously there would be an incentive for utilities to switch to cleaner and more renewable sources of energy. There would also be a strong incentive for households to reduce wasteful energy use.

The tax also creates an incentive to drive less and buy more fuel-efficient vehicles. Consider a driver who averages 15,000 miles a year in a vehicle that gets 17 miles per gallon. The 882 gallons of gas she uses would cost an extra $132.35 in the first year. But someone who limits themselves to 10,000 miles of driving in a year and has a car that gets twice those miles per gallon (34 mpg) would pay only $44.12 in tax. Of course, by the tenth year the tax is ten times higher, so the gas-guzzling road warrior is paying $1,323.50 in tax per year while her counterpart would be paying only $441 — provided that neither had changed vehicles or driving habits. On average, a household uses 1,143 gallons of gasoline a year.[59] That translates into a tax for the average household of $171.45 in the first year. If the household makes no changes by year ten, their tax would rise to $1,714.50 a year.

If you think that these tax figures are too high, or that they might cause hardship for poorer families, consider this: most Americans will be able to reduce the amount of tax they pay simply by conserving energy. And by doing so, they will not only spare themselves the tax, but the cost of the additional energy that they are no longer using. Much of this conservation can even be accomplished without investing in new technologies or energy conservation materials. Simply turning off unused lights, lowering the thermostat when not at home, making sure your car tires are fully inflated, or adjusting driving techniques could save the average American measurable amounts of money that they pay for electricity, natural gas, and gasoline. But most people won't take those steps until prices are high enough to persuade them to do so. In less than a decade, this BTU tax would provide a strong incentive for massive energy savings.

Of course, the direct taxes that Americans pay for the energy they use won't be the only cost increase that people see. Businesses that use energy to produce the goods and services that they sell will no doubt include some of that energy cost in the prices they charge. This shouldn't, however, have a major effect on prices, as energy is a relatively modest factor of production (compared to labor, land, capital, etc.) The Clinton Administration estimated that their tax, when fully phased in, would increase manufacturing costs on average by .1 percent. Since this tax in the first year is roughly twice Clinton's full BTU tax, it would raise manufacturing costs on average by .2 percent. If manufacturers did nothing to adjust their energy use (highly unlikely) the effect on production costs by the tenth year would be 2 percent.

But it's wrong to think that the end consumer will end up with the final bill for the entire BTU tax. Any time an item is taxed, the amount of the tax paid by consumers and the amount paid by producers depend on what economists call "the elasticity" of the supply and demand.[#] A 1993 report by the Federal Reserve Bank of Dallas noted that the BTU tax (proposed by Clinton) would cause the price of a barrel of oil, as seen by consumers, to rise, *but that the price received by oil producers would fall*. This meant that oil producers would have actually been paying part of the tax. According to the Fed's analysis, two-thirds of the tax on oil would be picked up by American consumers, while the other third would be paid for by producers. If you had the option of having the Saudis, Iranians, Kuwaitis, and Venezuelans (or even Exxon and BP) pay a

---

[#] "Elasticity" describes the responsiveness of demand (or supply) to a change in the price for that good or service. In other words, some goods can experience a huge increase in price and little or no change in demand, while others experience a great drop in demand with a relatively small increase in price.

third of your taxes, would you do it? Unfortunately, our Congress said no.

Besides making America more efficient, reducing our dependence on foreign oil, and improving the environment, there is one more very important benefit we must acknowledge. When energy is cheap, there is a greater incentive for businesses to automate production. As mentioned in the last chapter, this has been one of the two major job killers for the U.S. economy. Driving up the cost of energy reduces this incentive and, to some degree, protects American workers from losing their jobs to energy-intensive automation.

Measuring the Opposition

Just as in 1993, there is bound to be strong opposition to this kind of tax. Companies that make their living from extracting fossil fuels will fight to have the measure defeated, because they know it will put a serious dent in their long-term profits. Manufacturers — knowing they won't be able to pass on the entire increase in production costs — will claim that it will destroy jobs and give a competitive advantage to foreign competitors. These arguments might have merit if this BTU tax proposal weren't combined with stimulative economic measures and a tariff system that eliminates the production advantage of less developed countries. Our industrialized competitors also tend to have energy prices higher than our own — they've already realized the wisdom in taxing energy use — so this also isn't a problem. Industry spokespeople will no doubt widely proclaim the point that these taxes will cost consumers money. That's true. But the present system already costs us money — higher taxes (to pay for government programs and regulations), higher medical bills and health insurance (to combat the effects of pollution on the health of humans), decreased productivity (also related to health issues), and a weaker dollar (due to the portion of the trade deficit caused by higher imports of energy sources). At least under this proposed system, those who cause the externalities will pay for them.

Some will argue that this type of tax is regressive, since the poor and middle class pay the same rate as the wealthy. That's a valid point, but other parts of this economic plan make the argument moot. To begin, the income tax system is made much more progressive, shifting most of the burden onto the ultra-rich. The plan also completely eliminates the income tax for lower wage Americans *and* provides upward pressure on their incomes through an increase in the minimum wage. As pointed out before, even at the lowest end of the pay scale, the yearly scheduled increases in the minimum wage are easily enough to cover these taxes and still help lift the living standard of the working poor.

But what about those who are not working; isn't this tax unfair to them? The answer is no and yes. All Americans benefit from the services provided by the federal government, and should therefore have to contribute something in the form of taxes.[*] Under the income tax, non-workers don't have to pay a federal tax. For those who do not have employment, the BTU provides two important incentives. The first is to find employment, so they can afford the increased cost of energy and other products. A resource tax would be difficult to completely avoid, so it makes earning an income all the more important. The second is an increased incentive to reduce their tax burden by reducing their fossil fuel use as much as possible. While this reduction might limit their tax burden, it increases a benefit to society in general from the reduction in externalities. However, there are some — the disabled, and those on fixed incomes, like Social Security — who will be unfairly affected. While they too will have an increased incentive to reduce their fossil fuel use, some adjustments to government assistance programs that benefit these individuals may be required. The overall increase in government outlays should be minimal, though, particularly when compared to the economic improvements produced by this entire economic plan.

Of course, the fact that people will naturally be trying to avoid the tax by using resources more efficiently will cause some to make the argument that this form of government revenue is unreliable. In other words, as our country becomes more and more efficient, less and less tax will be collected. This is a possibility, but not in the near future. Since the BTU tax rate will be increased each year for the next ten, it is unlikely that we would see a decline in the overall revenue produced each year. If however, revenues declined

---

[*] It should be noted that this tax will be paid even by those who get their income from illegal or "underground" sources – something that can't be said of the income tax.

precipitously after the 10 year implementation period — in the odd case that we somehow eliminated our need for petroleum, coal, or natural gas — there would still be better alternatives for creating revenue than to re-implement a broad-based income tax. Remember, when the chapter began, I noted that the depletion of any natural resource threatens the existence of the economy. Although our immediate problem involves energy resources, we could eventually face the same problem with "non-energy" resources. So a reasonable argument could be made for putting some sort of tax on "non-energy" resources as well. While I'm not going to suggest that option at this time, if the revenue from an energy-based tax dwindled enough at some point in the future, a "non-energy" natural resources tax would be a wise, available alternative.

This non-energy natural resource tax would encourage a more efficient use of the elements used in creating steel, aluminum, glass, and even paper. Maybe more importantly, it would encourage the recycling of previously used materials, and reduce the amounts simply discarded in landfills. Take glass, for instance. It ought to be the perfect commodity to recycle. Glass can be recycled an infinite number of times, and is done so in a process that isn't particularly complicated. And yet there is an overabundance of collected glass lying around in various places, waiting to be recycled. This is because the main ingredient in glass, sand, is so plentiful. In fact, sand is so plentiful that one might think we would never run out. That is exactly what they thought at one time about petroleum. Even if we were guaranteed to never run out, does it make sense to melt sand into glass, use it once, and then chuck it into a landfill? And yet tons of our country's natural resources — not just sand — are used in exactly this way.

Think of how little trash you'd see on the street if the price of paper, aluminum, glass, and tin were driven up because their component materials were taxed. Recycling centers would offer a greater price per pound because getting recycled materials would be cheaper than pulling raw materials, which would be taxed, from the earth. Cities would institute recycling programs alongside their trash collection, not because of a federal mandate or lobbying by environmentalists, but because collecting and recycling these materials would be another source of revenue for the city. And because the tax would largely be passed on to consumers, this would encourage saving versus consumption, much the way a sales tax, or a VAT (value added tax) would.

Regardless of what, or whom, is taxed in America, there will always be some sort of complaint, accompanied by arguments against the tax. The goal of our government shouldn't be to protect or target a group of people, per se, but to establish a tax system that reduces the activities that cause problems for our society and provides a foundation for long-term economic growth. All taxes proposed in this plan, including a BTU tax (and possibly at some point a "non-energy" natural resources tax) do just that.

Conclusion

The future strength of America is closely tied to the way we approach our use of natural resources. Because of the limited nature of these assets (and in some cases their dwindling supply) the economy of the future — not just in the United States, but around the world — revolves around using natural resources, particularly energy resources, more efficiently.

Other countries have realized this and are aggressively pursuing energy efficiency programs as well as developing businesses that either produce alternative forms of energy or help businesses and citizens use present sources more efficiently. In 2009 China spent almost twice as much as the United States on clean energy investments, despite having a smaller economy. And for the last five years, our rate of clean energy investment growth has lagged behind five other G-20 members (Turkey, Brazil, China, the U.K., and Italy).[60] As people realize the cost savings that can be obtained from energy efficiency and the benefits of using alternative energies, businesses that provide these products will be in high demand. If America fails to encourage the domestic development of these industries in some way, it will simply be one more arena in which we have fallen behind the rest of the world and lose jobs to foreign competition. And, if we end up wasting our natural resources while the rest of the world is saving theirs, future generations of Americans will have to steal, beg or fight for the resources they need to keep the economy functioning.

We can avoid this unnecessary tragedy. If we want to gain a competitive edge in future industries such as alternative energy and energy efficiency; if we want to establish an economy that will survive the depletion of fossil fuels; if we want to wean ourselves from foreign oil; if we want a healthier society for ourselves, our children, and our children's children; we must address the way that we use our finite resources.

And while there are numerous ways to do this, the best way is through prices: driving up the prices of our limited resources so they more accurately reflect the cost. Prices are the perfect stimulant for driving change in the marketplace. Nothing motivates consumers to change their habits more than prices. Businesses use price changes as a signal to create new products and services, as well as to change processes and business models. Instead of having subsidies and tax credits for making energy efficiency upgrades, people and businesses would make these changes on their own in an attempt to save money on their power and heating bills. Instead of creating government programs to help develop alternative energy sources, the private sector would create these alternatives because renewable energy sources like wind and solar would suddenly be cost competitive. This system would substitute bureaucratic regulation with market forces that drive us in the same direction. Instead of creating costly programs that use up government revenues trying to find the best answers, we'd use the power of the market to develop the best solutions and at the same time increase government revenues and help to balance the budget.

The Federal Budget

Let's stop at this point and see what we have accomplished, assuming that the plan detailed up to this point has been enacted. We have revived the economy, not through budget-busting tax cuts or big government spending proposals, but by insuring that the income of our nation is distributed in a more equitable manner. There is still income disparity; there are still incentives to work hard, create small businesses or climb the corporate ladder; there are still rich individuals and families, as well as poor individuals and families, but the gap between the top and the bottom isn't so wide that it threatens the proper functioning of our economy. The economic environment for small businesses is better because people actually have the income to patronize them, and investment in new enterprises is encouraged through the tax code. We have put in place tax systems that make sense — low taxes on work; high taxes on income that stems from greed and market imperfections; tariffs that will halt the exporting of American jobs while not impeding legitimate international trade; and taxes on energy that will drive efficiency and protect the environment. Thus we have improved the outlook for our economy, our national security, *and* the environment.

Because our actions to this point have accomplished so much without putting pressure on the nation's finances, we can now focus on balancing the budget. To that end, we should note that our changes to this point have not only solved many of our problems, but have actually added additional revenue to government coffers. In comparison to figures put out by the Congressional Budget Office (CBO) in early 2011,[61] the system outlined in the previous chapters would bring in $3.9 trillion more in tax revenue over the next 10 years[*] than the currently legislated system, reducing projected budget deficits by that same amount.

Ultimately, however, that is not enough. Despite the additional revenue created by the plan outlined so far, our government's budget will still not be balanced at any point over the next decade. In fact, the same 2011 report by the CBO estimated that the United States will run budget deficits of over $6.9 trillion during that period, sending our national debt north of $20 trillion.[62] Because our additional revenue leaves us $3 trillion short of simply achieving balance, we will need to cut spending to make up the difference (and eventually produce a surplus so that we can begin to pay off the more than $14 trillion that we already owe). Each of these actions — balancing the budget and doing so by reducing government outlays — is important for a number of reasons.

Large, persistent budget deficits threaten our economy by driving up interest rates.[+] If the U.S. government's creditors begin to feel that our debts are unmanageable, interest rates could easily rise to levels that would put a serious crimp in any kind of economic revival that we have generated. In addition to the interest rate problem, the United States has become dependent on other nations for the financing of our debt. Currently almost one-third of America's national debt is held by foreign entities, giving them a dangerous and undue influence over our economy.

Our economy has also become too accustomed to having a certain amount of government spending each year. Even before the large spending initiatives enacted by the Obama Administration, our

---

[*] FY2012 - 2021

[+] As mentioned before, this phenomenon may be delayed if the money supply is expanding, which has been happening in the last few decades. But even an expanding money supply can't keep interest rates at bay forever.

government's outlays were much higher than they should have been, as evidenced by the trillions of dollars in national debt passed on from prior administrations. This debt represents how much our past and present standards of living have been raised above what we could afford, and how much (plus interest) future living standards will have to suffer in order to pay it back. Passing this type of debt burden on to our children and grandchildren is morally wrong. It is not right for one generation to improve its standard of living at the expense of another.

In addition, these debts and deficits cause us to lose a lot of valuable government tax revenue to interest payments. It is estimated that the interest payments of the federal government for the next 10 years will cost tax payers almost $5.5 trillion.[*][63] That's money that could easily be spent on infrastructure, education, or tax cuts. And once again, since almost one-third of the debt is held by foreigners, a lot of this interest is simply money that we are giving away to citizens of other countries, instead of keeping it here in the United States.

Finally, any estimate of future government revenues is bound to be different from what will eventually materialize. So even if our forecast showed a balanced budget, it would be better to be conservative and create some "wiggle room" by suggesting budget cuts, in case our estimated revenue doesn't come to fruition. If, as time passes, it turns out that the federal budget was cut too much in certain areas, or that government revenues are better than expected, our political leaders will certainly find it easier to increase outlays than to increase taxes or cut additional spending to make up a shortfall.

Many of the suggestions you might hear in the public arena — eliminating earmarks, cutting education, eliminating the EPA, halting foreign aid, etc. — simply won't provide enough savings to balance the budget. Not that cuts couldn't or shouldn't be made in some of those areas, but expecting to bring about a budget surplus by focusing on these relatively minor budget items is futile. To demonstrate my point, the following departments could have been

---

[*] This is for fiscal years 2012 – 2021 and is a net interest figure. The government will pay more than this in interest, but some of that will come back to taxpayers through debt obligations held by other government agencies.

completely eliminated from the federal budget for the last 20 years without eliminating the yearly budget deficit:

>Army Corps of Engineers
>National Science Foundation
>Environmental Protection Agency
>General Services Administration
>NASA
>Department of Commerce
>The Judicial Branch (one of the 3 branches of our government!)
>Housing and Urban Development
>Department of Education
>Department of the Interior

The only exceptions to this are fiscal years 1997 to 2001 when the federal budget was relatively in balance. And I don't mean that any *one* of these departments could have been eliminated, I mean that *all of them* could have been dropped without producing a balanced budget. Once again, this isn't a statement about the legitimacy or worth of these departments or programs, or even a statement about the political feasibility of eliminating any one of them, let alone *all* of them. My point is simply that if you're serious about balancing the budget, you have to address the major parts of the budget, not just eliminate portions that may not be popular with the particular political base you're trying to win over.

Also in the heap of poorly developed ideas for reducing spending is the "cut everything in the budget, except defense, by X percent" plan, with X representing some amount that is large enough to sound impressive, but not large enough to sound radically scary. I've seen both Democrats and Republicans suggest this approach, similar to its close cousin, the "freeze all spending, except defense, at such-and-such a level for X number of years" approach. These proposals not only display laziness in regard to rooting out real waste and unnecessary spending, they assume that the voting public can't handle the complexity and somberness of an in-depth discussion about the nation's finances. Some departments and spending programs in the federal budget — those that continually seem to be on the chopping block — are much more efficient than others. Using the same "cookie-cutter" approach on all of them is a poor way to bring balance to the budget.

While this type of approach may appeal to some constituencies, one wonders if those citizens would still be enthusiastic about such a proposal when they found out that those cuts meant fewer food safety inspections or less air traffic control. These proposals also usually exempt one of the most bloated, pork-laden parts of federal spending — the Defense Department. Roughly half of all earmarks end up going into the defense budget.[64] So if you're going to eliminate the wasteful earmark spending that our politicians continually demonize, you've got to make cuts to the defense budget.

In reality, there's a very uncomfortable fact that Americans need to face. If we really want to get serious about balancing the budget, we have one of four options:

1. Increase Tax Revenues
2. Cut the Defense Budget
3. Cut spending on Medicare/Medicaid
4. Some combination of the first three (the approach outlined in this book)

The reason is simple. If we look at "on-budget" revenues and expenditures, the first three items listed above are the major factors affecting whether or not the budget is in balance.[*] In 2010, defense and Medicare/Medicaid expenditures accounted for 49 percent of "on-budget" expenditures.[+] If you include net interest on the national

---

[*] Interest on the national debt is also a major factor, but failing to pay that interest would throw the country into economic chaos, and therefore isn't considered here. The section of the budget known as "Income Security" — things like unemployment benefits, housing and nutrition assistance — is also a major segment of the budget, but hasn't been growing as a percentage of government expenditures until this latest recession, whereas defense and Medicare/Medicaid have both been consuming greater and greater portions each year. Additionally, as the economy recovers and the effects of this economic plan take hold, "Income Security" expenditures will naturally decrease. The same cannot be said for Medicare and defense spending.

[+] The Medicare system receives "off-setting receipts" in the form of Medicare premiums from beneficiaries. These have been stripped out of Medicare spending for this calculation to give a more honest assessment of the impact of these programs on our government's finances.

debt, that percentage climbs to over 56 percent. If you compare these expenditure items to the general revenue (which pays for "on-budget" items), defense/Medicare/Medicaid took up 94 percent of that revenue, and net interest on the national debt gobbled up 13 percent! In other words, what we're currently collecting in taxes barely covers these three budget items: defense, Medicare and Medicaid. Throw in interest on our debt and we're already in a deficit position. And this situation only gets worse in the future as interest payments on the national debt grow. So if you're serious about balancing the budget (and paying down the national debt), you either have to raise more revenue, cut spending in the major segments of the budget, or both.

While bringing "on-budget" accounting into balance will be a tremendous step, it will do little good if we don't also solve the long-term fiscal gap in the Social Security system. Although not in "crisis," as some have claimed, Social Security does have a serious long-term fiscal imbalance. Therefore this chapter will conclude with recommendations for bringing long-term solvency to that system.

Addressing the Cost of Health Care

Much of the fiscal problem with both Medicare and Medicaid revolves around the ever-increasing cost of health care. In order to bring both of these programs under control, we must first address the rapid rise in healthcare costs. While it has been suggested that the recently passed Patient Protection and Affordable Care Act will help to bring down future medical costs, it's obvious that this plan was not comprehensive and more will need to be done. This will be no easy task.

The universal question that emerges in any debate about health care is whether or not health care is a right of every American. While it's a worthwhile question, it avoids a very important responsibility surrounding the rights and freedoms we enjoy. Any right or freedom that we have, as Americans, is accompanied by a duty to value that right, or use that freedom responsibly. If, indeed, we each have a right to health care, then we also have a responsibility, as individuals and as a nation, to keep ourselves as healthy as possible. This seems to be forgotten in the healthcare debate. Hence we have a situation where a lot of individuals fail to take care of their own health, the government fails to protect us against public threats to our health, and everyone thinks the solution is more healthcare services. Not only is this reactive approach foolish and irresponsible, it increases the demand and price of healthcare services for everyone.

While I'm not an expert in healthcare spending, it would seem obvious that adhering to the following guidelines would make a world of difference in reducing prices in the healthcare arena:

1. **Implement policies that promote a healthy living environment.** Clean air, water and soil are vital for humans to remain healthy. The more polluted our environment becomes, the less healthy the general population will be. To the degree that pollutants increase the incidence of everything from asthma to cancer, they drive up the demand, and cost, of health care. Therefore, promoting policies that reduce the damage caused to our environment are vitally important. The BTU tax proposed in the last chapter is a major step in the right direction, but more could be done. The government must be strict about enforcing the amount of contaminants allowed under our country's environmental regulations. If pollution causes higher rates of disease and those higher rates drive up healthcare costs, shouldn't the polluters have to pay the additional healthcare expenses? We have a system that makes our citizens pay (in one way or another) while polluters get off with a slap on the wrist, or allows them to tie up the court system fighting their penalty until the government ultimately settles for a reduced fee. That system must somehow be rectified.
2. **Implement policies that promote healthy living.** People complain about "sin taxes," but they make absolute sense. A price needs to be paid for engaging in activities that increase a person's chance of getting various diseases. Taxes on tobacco products, alcohol, even fast food and junk food, ensure that the people ingesting these substances are paying that price.
3. **The lowest paid full-time employee must be able to afford health insurance.** The guidelines mentioned in this list will mean little if workers can't afford their own health insurance. If the typical person is taking the appropriate steps to maintain his or her health, all that is

needed is a low cost, high deductible health insurance plan that would cover unforeseen accidents and enough money to cover yearly preventive care needs. If the minimum wage isn't high enough to provide that, then the system immediately begins to fail. People begin looking for government assistance to obtain health care, and a logical argument is born for providing that assistance, because even those who are working can't provide it for themselves. Next, the market becomes distorted due to government interference, appropriate incentives are destroyed, and the downward spiral of failing health and out-of-control prices begins. Reducing income inequality and insisting on personal responsibility are absolute necessities for instituting a healthcare system that functions logically.

4. **Make sure people recognize the cost of their health care.** An alternative to the "sin taxes" mentioned above would be to make each individual responsible for purchasing his/her own health insurance. This would mean abolishing not only Medicare and Medicaid, but also the extremely popular employer-provided health coverage. If individuals and families were responsible for purchasing their own insurance or health care, insurers and healthcare providers would make sure to factor each buyer's personal habits and activities into what they were charged. This would deliver a price signal, similar to that of a sin tax, that certain unhealthy activities should be avoided.

5. **We have to come to terms with the fact that we all die (and death isn't the worst thing).** Our society seems to have lost its ability to cope with death. It's evidenced in the fact that we spend a disproportionate amount of our lifetime health expenditures in the final months of our lives. Much of this near-end outlay is spent simply trying to prolong life for another couple of weeks or months. While the desire to live as long as possible is certainly understandable, it must be weighed against the cost of prolonging life for what is ultimately a relatively small amount of time. Once again, this goes back to people seeing the true cost of their health care. If you knew that

you could prolong your life by one more week, but the medical expenditures required to do so would saddle your surviving family with debt for the rest of their lives, would you do it? Most people would probably say "no," but under the present system, too many people are deciding to make those medical expenditures because their family doesn't see a lot of the bill — it is paid for by society through Medicare, Medicaid, or higher health insurance premiums on others.

6. **Medical malpractice reform.** There needs to be a system in place where those injured through medical malpractice can find a remedy for the injustice done to them. But this system must also be one that limits frivolous lawsuits brought by individuals simply looking to get rich quick. To some degree, this problem has already been addressed by this economic proposal. Since yearly income in excess of $900,000 is taxed at such a high rate, there is little incentive to seek legal settlements in excess of that amount, unless the plaintiff truly wants to punish the medical professional who committed the malpractice. Just as the suggested income tax system deters market actions motivated by greed, it deters lawsuits with the same motives.

7. **Support preventive health care without socializing the cost of avoidable procedures.** It makes sense to have a system that encourages preventive care (even if it means artificially lowering the prices for those services), because it will help individuals avoid more costly procedures in the future. While this might seem contrary to some of the guidelines mentioned above, remember that there is a social benefit to avoiding preventable illnesses, making a system that rewards steps taken to ensure that this prevention occurs, sensible. The important distinction is that this system should not also be one that helps pay for the treatment of illnesses that are the result of someone's personal choices or habits (smoking, drinking excessively, eating poorly, etc.).

8. **People have to be more important than profits.** Up until 1997, pharmaceutical companies were essentially unable to advertise prescription drugs on television. This

was due to the fact that any drug advertisement had to list a brief summary of the precautions, warnings, contraindications and adverse effects surrounding the drug's use — something that was practically impossible in brief radio and TV spots. In August of that year, the Food and Drug Administration, after much lobbying by the pharmaceutical industry, relaxed the guidelines for broadcast advertising, essentially opening the television airwaves to drug commercials. Since that time, the pharmaceutical industry has done a great job of convincing Americans that there is a pill to solve our every problem. They've also done a great job of driving up the cost of health care and creating immense profits for their industry. The pharmaceutical industry now spends roughly twice as much on advertising as it does on research and development,[65] and Americans have become accustomed to using high-priced pharmaceutical drugs to battle their illnesses instead of cheaper, more natural solutions that have fewer, if any, side-effects. This is a perfect example of what happens when the government puts the profits of the healthcare industry before the welfare of the people.

While this probably isn't an exhaustive list, using these concepts as a basis for making changes in healthcare legislation would make a world of difference. Some of these guidelines have already been addressed, to some degree, by the changes suggested earlier in this economic plan. For instance, the BTU tax placed on energy should not only create a healthier living environment by reducing air pollution, it should encourage healthier living by increasing the incentive to walk or cycle from one point to another. Some of the guidelines may not be addressable by legislation at all; they simply require us to get over our feeling of entitlement toward unlimited health care. If, though, we could stick to these principles when it comes to making healthcare decisions, we might finally start to see the nation's healthcare cost curve begin to bend downward.

Medicare

We can't simply rely on uncertain healthcare cost savings to reduce our expenditures on programs like Medicare. We must make adjustments to the program that *guarantee* savings in coming years. Before we get into those modifications, let's take a brief look at the Medicare system.

Medicare is broken up into numerous parts (A, B, C, and D). Part A pays expenses for inpatient hospital stays, skilled nursing facilities, and certain home health care services. Part A is financed primarily through a 2.9 percent payroll tax on all employees (1.45 percent paid by the employee, 1.45 percent paid by the employer). Part B covers physician services, outpatient hospital services, certain home health care services, and durable medical equipment. Part B is financed through general revenue from the U.S. Treasury (i.e., tax dollars) and a monthly premium paid by all beneficiaries. Part C covers all expenses covered by parts A and B and is provided by government-approved private health insurance plans. Depending on the insurance plan, other benefits and prescription drug coverage may be included. Part D provides prescription drug coverage. Like Part B, Parts C and D are funded by general tax revenue and premiums paid by the enrollee.

When Medicare was established in 1965, the eligibility age was set at 65 and the average life expectancy was 70.4, so people weren't expected to be drawing benefits from the program for very long. Since that time, the average life expectancy has risen to over 78 years of age, but the eligibility age hasn't changed. Therefore the average person now has access to Medicare benefits for a longer period than when the program originated. These additional years are during a period of life when healthcare expenditures are highest.

So the first adjustment that should be made to the program is to begin to increase the eligibility age for receiving benefits. This is a logical modification given the increase in life expectancy. Under this plan, the eligibility age for Medicare benefits would slowly be increased according to the schedule on the following page. While this timetable begins sooner than other proposals put forward for increasing Medicare's eligibility age, it makes the adjustments more gradual — increasing the eligibility age by only one month per year for the first 24 years, rather than two months per year. It also increases the final eligibility age for those born after 1981 more than other proposals, which typically stop at age 67. The budgetary impact of this change in the next decade is minimal, but over the long haul it will be substantial and will go a long way to making the Medicare system financially sustainable.

Medicare Eligibility Age

Birth Year	Current Eligibility Age	New Eligibility Age	# of additional months a retiree must work:	
1949	65	65	0	more months
1950	65	65	0	more months
1951	65	65	0	more months
1952	65	65 and 1 month	1	more month
1953	65	65 and 2 months	2	more months
1954	65	65 and 3 months	3	more months
1955	65	65 and 4 months	4	more months
1956	65	65 and 5 months	5	more months
1957	65	65 and 6 months	6	more months
1958	65	65 and 7 months	7	more months
1959	65	65 and 8 months	8	more months
1960	65	65 and 9 months	9	more months
1961	65	65 and 10 months	10	more months
1962	65	65 and 11 months	11	more months
1963	65	66	12	more months
1964	65	66 and 1 month	13	more months
1965	65	66 and 2 months	14	more months
1966	65	66 and 3 months	15	more months
1967	65	66 and 4 months	16	more months
1968	65	66 and 5 months	17	more months
1969	65	66 and 6 months	18	more months
1970	65	66 and 7 months	19	more months
1971	65	66 and 8 months	20	more months
1972	65	66 and 9 months	21	more months
1973	65	66 and 10 months	22	more months
1974	65	66 and 11 months	23	more months
1975	65	67	24	more months
1976	65	67 and 2 months	26	more months
1977	65	67 and 4 months	28	more months
1978	65	67 and 6 months	30	more months
1979	65	67 and 8 months	32	more months
1980	65	67 and 10 months	34	more months
1981	65	68	36	more months
1982	65	68	36	more months

The second change that should be made is to reduce the income thresholds at which Medicare recipients start to pay more of their Part B premium. The full Part B premium for a Medicare recipient is roughly $440 per month (2010). For most recipients, the government has always paid 75 percent of that premium. But in 2007, beneficiaries with higher incomes started paying for more of their Part B premium, as mandated by changes to the program in 2003.

Unfortunately, the income thresholds were set unusually high. No rate increase is seen until an individual's Modified Adjusted Gross Income[*] (reported to the IRS) surpasses $85,000 a year ($170,000 for couples). Even then, the government is still picking up 65 percent of the tab until an individual's income surpasses $107,000 ($214,000 for couples), at which point the beneficiary pays 50 percent of the premium and the government pays the other 50. Premium adjustments happen again at income thresholds of $160,000 and $214,000 for individual filers ($320,000 and $428,000 for couples filing jointly) above which the beneficiary is paying 65 percent and 80 percent of the premium, respectively.

A couple of points should be stressed here. The first is that Part B is funded strictly through the minimal premium payments mentioned above and general tax revenue. At no point during a beneficiary's life did he or she pay a special tax to fund this part of the Medicare system. It is not as if the government *owes* retirees this benefit because of previous contributions they made. The second point is that we are talking about *retirement income*. If your *retirement income* is in the hundreds of thousands of dollars per year, the American taxpayer shouldn't be subsidizing your health insurance. Most Americans don't earn $100,000 a year during their working lives. Why should we tax them to provide a benefit to those who make that much during retirement? Medicare should be a safety net for the elderly who are poor, not a hammock for the rich.

For this reason the income thresholds should immediately be adjusted downward. Starting in 2012, any individual with a Modified Adjusted Gross Income between $50,000 and $75,000 a year ($100,000 and $150,000 for couples) should pay 50 percent of the

---

[*] Modified Adjusted Gross Income is a retiree's Adjusted Gross Income plus their tax-exempt interest income.

Medicare Part B premium ($221 a month in 2010). For individuals earning between $75,000 and $100,000 ($150,000 and $200,000 for couples), the premium should be 75 percent of the total. Anyone making more than $100,000 a year in retirement ($200,000 for couples) should pay the full Part B premium. Even the premiums required under these new income thresholds are somewhat generous to wealthier, older Americans, so it's debatable whether or not these levels should even be adjusted for inflation on a yearly basis (at least for the first five or ten years after the change). Obviously, without the inflation adjustment, more and more retirees would theoretically be picking up the entire cost of their Part B Premium as time passed. Over the next ten years, without an adjustment for inflation, this change should cut Medicare expenditures by approximately $121 billion.

These same income thresholds should also be established for the Medicare Part D program. Currently higher income individuals pay their subsidized drug-coverage premium, plus an additional amount equal to 35, 50, 65, or 80 percent of the total cost[*] on the basis of the income thresholds mentioned earlier. Under this arrangement, the new income groups listed above would pay their subsidized premiums, plus an additional percentage of the base premium in line with the percentages mentioned above (50, 75, and 100 percent). Once again, these income levels would not be adjusted for inflation in the first five to ten years, so the number of beneficiaries affected by the higher rates would theoretically grow every year. The 10-year budget savings from this modification would be sizable, although not as great as the $121 billion gained from the adjustment made to Part B.

The final change we should make to the Medicare program is to allow the administrators of the program to bargain with pharmaceutical companies on drug prices, the same way that the Veterans Administration (VA) does. This type of negotiation was explicitly prohibited by the legislation that created the program — an obvious gift to the pharmaceutical industry — but could provide huge savings to the Medicare program. In fact, it is estimated that this change would provide $14 billion worth of savings to the

---

[*] Because Medicare drug premiums vary, the additional amount paid is calculated using a base beneficiary premium.

Medicare program each year. Of course, one of the ways the VA increases its negotiating clout is to restrict the number of drugs it covers in each drug class. If the Medicare program adopted this tactic, it would limit the choice of beneficiaries to some degree for certain classes of drugs. This loss of choice could easily be made up for by taking half of the $14 billion savings per year and using it to reduce premiums — something that would probably be more meaningful to beneficiaries than a greater choice of drugs — while still providing $70 billion worth of savings to the taxpayers over the next ten years.

Medicaid

While Medicare is meant to provide a minimum level of health care for older Americans, Medicaid is meant to be a healthcare safety net for poorer Americans. It is a partnership program between the federal government and the states that helps provide health care for America's needy. The federal government has set minimum guidelines for who is eligible and what benefits should be offered. The states have to meet those guidelines, but can then go further, if they desire, by loosening the guidelines and expanding the range of benefits. The federal government reimburses states for their Medicaid expenditures at a predetermined rate, different for each state.

Because Medicaid is an assistance program for the poor, the economic changes outlined in the last few chapters will help reduce Medicaid expenditures in a couple of ways. A greatly improved economic picture in the United States will reduce the unemployment rate and, in turn, reduce the number of people relying on Medicaid. Reliance on Medicaid will also decrease because incomes will be raised at the lower end of the income scale, making fewer of the working poor eligible for the benefits. Instead of receiving a government handout, more people will have an income that enables them to be responsible for their own healthcare coverage.

But, as with Medicare, there are problems with the way the system has been set up. Therefore, we can do more than simply rely on a good economy or reduced healthcare costs to produce savings in the system; we can change the structure of the system to *guarantee* savings.

The change that gets mentioned most often when it comes to reforming the Medicaid system is to replace the federal reimbursement system with one that uses block grants. As mentioned earlier, the federal government reimburses states for a portion of their Medicaid payments. While the reimbursement rate varies from state to state, the federal average is roughly 57 percent.*[66] In other words, on average, each time a state spends one dollar on Medicaid, it receives $1.33 from the federal government.+ This creates a perverse incentive for states to go beyond the minimum federal coverage requirements and create rather generous healthcare programs for the poor. Block grants would solve this problem by setting a fixed amount that each state would receive from the federal government regardless of what it spends.

The problem with this idea, however, is that it eliminates the incentive for states to spend *any* money on their Medicaid programs. Why would a state spend its own money if it was guaranteed a set amount from the federal government? In an era of tight state budgets, a block grant system might cause an overly dramatic cut in programs for the poor.

A better solution would be to slowly reduce the reimbursement rate that Washington gives to each state so that total spending on the program by the federal government grew no faster than a certain percentage each year. For instance, starting in 2013, the growth rate of federal spending on Medicaid could be limited to 6 percent a year. Reimbursement rates to all states would be adjusted downward to stay within this total growth rate limit of federal spending. If eligibility rates were low enough in a given year that the 6 percent growth was unnecessary, this would provide extra savings, but the system would guarantee that federal outlays for Medicaid don't grow out of control. In fact, limiting the growth of federal Medicaid spending to 6 percent a year would save over $700 billion dollars over the next decade.

Of course the impact of this change would be different in every state, because each state has control over what benefits it

---

* This average percentage was increased during the latest economic downturn, but is expected to return to 57%.

+ If the math here seems odd, consider a state that establishes a $1 million Medicaid program. If the state is receiving a 57% reimbursement from the federal government, the state will spend $430,000 and receive $570,000 from the feds to pay for the program. So Uncle Sam is giving the state $1.33 for every $1 it invested.

offers and what income levels are eligible, as long as it is meeting the federal minimum. Some states might eliminate eligibility for those with higher incomes.* Some states could limit the number of childbirths per mother that are eligible for assistance. (If someone is too poor to provide health care for her children, should a state be subsidizing an unlimited number of deliveries for that person?) It may even be possible that the economic changes that develop from the rest of this plan reduce the need for Medicaid spending without any adjustments being made by the states. But we can't simply *assume* that this will be the case. In order to get the budget under control we must put modifications in place that will guarantee a certain amount of savings.

Establishing a Real Defense
If you bought a home security system for $10,000, and then someone broke into your house with a screwdriver and killed your family, would you think, "Gee, I should probably spend another $1,000 upgrading my security system"? No, you'd wonder why the first $10,000 you spent wasn't enough to stop the intruder to begin with! Why would you spend any more money upgrading a home security system that had already failed? Unfortunately, our political leaders have convinced us not to think that way.

On September 11, 2001, the United States experienced one of the worst attacks in our nation's history. Up until that time, America had been spending more on national defense than any country in the world. In fact, we were almost spending more than all other countries *combined*. And yet on that day, assailants — trained in one of the poorest countries on the planet — killed nearly 3,000 Americans and knocked down our country's two tallest buildings. We can say what we want about the style of attacks on September 11th or the people who carried them out, but if a dozen people can kill three thousand of our fellow citizens *on our own soil*, there's something wrong with our defense structure. And while we saw plenty of finger pointing in the months following the attack — Republicans blaming the Clinton Administration, Democrats blaming Bush, and Bush blaming the CIA and the FBI — no one

---

* Currently states are allowed to offer benefits to those making up to 185 percent of the federal poverty level.

suggested that maybe we had been misallocating the hundreds of billions of dollars we spend every year to protect ourselves. Instead, we decided to "buy more security" by creating the Department of Homeland Security.*

I bring up the attack of September 11 and our response to that event because it highlights some of the problems with our attitude and approach to our defense spending.

Our primary problem is that Americans have been led to believe that more military spending automatically translates into a better defense of the nation. Anyone who is not in favor of an ever-increasing military budget is branded as "unpatriotic" or "weak on defense." The attitude is ironic, because many of those pushing this notion are the same ones who tell us that problems in other departments of the government, such as education, won't be fixed "by simply throwing money at them." Somehow, when it comes to defense spending, more is always better, no matter where the money is directed.

This unfounded attitude toward our military spending has led to the second problem with our defense structure: our defense budget is too often used as a means for economic stimulus. The fact that such a large portion of earmark spending ends up in the defense budget is a sure sign that senators and representatives are using this portion of our government's outlays to bring money and jobs back to their districts. Whether or not this money represents a good investment for providing the common defense seems to take a back seat to helping entrenched political leaders win re-election by keeping unemployment low in their districts.

And this brings about the final problem with Pentagon spending: we often spend money on things we don't need, while neglecting to spend money on things we do. Had our defense spending been more focused on our necessary needs, we would have cut expenditures on "Cold War" armaments two decades ago when

---

* Notice that no one on Capitol Hill calls this new branch of government the Department of Homeland Defense. If they did, taxpayers might begin to wonder why we have to pay for a Department of Defense *and* a Department of Homeland Defense. This underscores the fact that the Department of Homeland Security is really just another "big government" solution – akin to creating a Department of Learning to make up for failures in the Department of Education.

the Soviet Union collapsed and the Berlin Wall fell. The savings could have been, at least in part, moved toward developing better intelligence capabilities and bolstering security systems here in America. While spending money in these areas doesn't engender the patriotic fervor that building tanks and aircraft carriers does, it would have done more to blunt the terrorism threat emerging in the early and mid-1990s. To make matters worse, the next threat — cyberwarfare — has already begun to emerge and the United States is still struggling to find ways to curb procurement spending on weapon systems we don't need.

Encouraging the misplaced belief that more military spending translates into better defense is a lobbying army that represents the corporations that make up the military industrial complex. The drive for greater profits by these corporations has not only fostered our attitude and approach toward defense spending, but our viewpoint of America's place and role on the world stage. Our constitutional obligation to "provide for the common defense" has morphed into one of "protecting American interests abroad." This latter goal requires a much more expansive military and worldwide presence, essentially making America the world's policeman. A perfect example of this is the fact that the United States feels obliged to get involved any time there is a problem in the Middle East, because we have failed to solve our energy problem domestically and now see Middle Eastern oil as an American interest. Not only does the idea of protecting American interests abroad have no basis in the Constitution, what might represent "American interests" to one citizen might hardly be worthwhile to another.

If the United States is truly committed to defending its citizens, our government must reject the influence of the military industrial complex and begin to focus on the true present and future threats to our nation. We must return to our true constitutional responsibility of providing for the common defense and reject the notion that we can solve our domestic problems through nation building and military actions abroad. Never again should we find ourselves in the circumstances we were in on the morning of September 11th, 2001 — with military personnel and equipment parked all over the world, but unable to protect our citizens here at home.

Re-establishing Our Defense

Impressive suggestions have been made recently for reducing unnecessary military expenditures over the next decade. These include submissions by The Task Force for a Unified Security Budget[67]; The Sustainable Defense Task Force[68], made up of Rep. Barney Frank (D-MA), Rep. Ron Paul (R-TX) and other reform-minded military analysts; Benjamin Friedman and Christopher Preble of the CATO institute[69]; and the Center for American Progress.[70] While the suggestions of each group vary in what is cut and in the timeframe for savings, each analysis shows that there are massive savings to be obtained from whittling the defense budget. This section borrows suggestions from these groups to come up with just over a trillion dollars of savings over the next decade. While the bulk of these savings would help to produce a balanced budget, some of the funds should be redirected into other areas, such as border security, first responders and other "homeland security" assets, and cyberwarfare defenses, where America must do more to mitigate current and future threats.

Cuts should begin with a reduction in America's nuclear arsenal. An article put out in the spring of 2010 argued that deterrence from nuclear war could be obtained by the United States with a nuclear arsenal of 311 warheads.[71] The new START treaty recently signed by President Obama calls for reducing the number of America's deployed warheads to 1550, so greater cuts could easily be made beyond those to which we've already agreed. Reducing our number of deployed warheads to 500 (with 50 additional in storage) could save the United States a conservatively estimated $100 billion over the next decade (this includes reductions in our number of Trident and Minuteman missiles, along with ending work on the Trident II missile). With the number of deployed warheads reduced, we could easily retire the bomber leg of our nuclear triad — the weakest of the three delivery options — and scale back the number of nuclear subs in operation by 50 percent (7 SSBNs instead of the currently planned 14). The savings obtained from reducing the number of active nuclear subs (including personnel and operations and maintenance [O&M]) would be in the ballpark of $5 billion over the next 10 years.

America should not simply make these reductions unilaterally, though. Requiring further disarmament by Russia in the process of this arms reduction would not only enhance security for the United States and the rest of the world, but also garner both countries greater respect on the international stage. Additional creativity in international negotiations might also bring about guarantees of nuclear nonproliferation from other countries in conjunction with this drawdown. If countries like Iran and North Korea could be brought into the negotiations, we might even be able to halt further development of their nuclear programs while saving money on our own.

It makes no sense to reduce our current nuclear arsenal while continuing to spend money on new nuclear facilities and development. In conjunction with our nuclear pullback, we should cancel the construction of three new facilities tied to the development of new nuclear armaments. Cancellation of these proposed facilities in Los Alamos, New Mexico; Oak Ridge, Tennessee; and Kansas City, Missouri, would save an estimated $6 billion over 10 years.[72]

During the Cold War, when America was maintaining a greater number of warheads, the Department of Energy spent $2 billion less per year than it does today.[73] An estimated $20 billion could be saved from 2012 to 2021 by implementing a more efficient approach to maintaining our stockpile of reliable nuclear warheads.

Whether or not spending on missile defense is worthwhile has been debatable for some time. The trillions of dollars that we've spent on nuclear weapons were supposed to act as a deterrent against a foreign strike by nuclear missiles. We have the capability to determine where warheads are launched from and an arsenal large enough to respond with a barrage of destruction that would make any nuclear attack against the United States foolish. Spending hundreds of billions of additional dollars on a system that will protect us against the same type of attack seems an obvious waste when our borders are porous enough for terrorists to enter the country without our detection. Such are the spending decisions made in a town dominated by lobbyists (not a few of whom work for the defense industry). If America is to be truly safe in the future, our leaders must make common sense decisions about defense spending instead of earning themselves campaign contributions from military contractors.

Given our already sizable investment in this questionable program, however, it makes sense to continue development on certain portions of the program, while at the same time realizing budget savings by curtailing others. The Congressional Budget Office estimates that $51 billion could be saved by eliminating the Missile Defense Space Experimentation Center, the Airborne Laser, the Space Tracking and Surveillance System, the Sensor Development program, Far-Term Sea-Base Terminal Defense, and "Special Programs," as well as halting new program development until current systems have been proven effective.[74] The Sustainable Defense Task Force also recommends truncating the Space-Based Infrared System (SBIRS) for an additional savings of $2.1 billion over 10 years.[75] Part of the reason for the SBIRS program was to detect short-range theater missiles like those faced by U.S. troops during the Persian Gulf War in 1991. Once again, if America focused more on national defense than "defending American interests" the need for such programs would be greatly reduced. Roughly $55 billion dollars could easily be cut from the missile defense program over the next ten years with no reduction in national security. And once again, these cuts might actually improve our relations with other nations, opening the door for greater cooperation on issues that *really* threaten international peace.

Troop strength is another area where reductions can be made and savings experienced. By reducing the presence of U.S. troops in Europe and Asia and rolling back the growth in Army and Marine Corps units due to the wars in Iraq and Afghanistan, end strength of military personnel can be cut by 174,000. These cuts could include 109,000 active Army soldiers, 37,000 Marines, 10,000 Air Force and 18,000 Naval personnel. While these cuts might seem large, they still allow our country to field a combined (Army and USMC) expeditionary ground force of over 400,000 active-duty troops, meaning we could place 140,000 soldiers overseas at any one time. Bolstered by reservists, a force of this size should easily be able to handle most foreign engagements, although it wouldn't be large enough to permit us to get bogged down in a pair of dual nation-building affairs as we have been for the last decade. In other words, while providing enough of a force to "provide for the common defense," these troop reductions wouldn't give our leaders the luxury of using our men and women in uniform as foolishly as they have in recent years. The savings would be well worth it: over the next decade, these cuts to troop strength could save the United States $225 billion.[76]

As the wars in Iraq and Afghanistan wind down, there should also be less of a need for the expanded recruitment efforts enacted over the last decade (recruitment costs have nearly doubled since 1999). Gradually reducing recruitment expenditures from 2012 to 2021 should save a total of $5 billion.[77]

When it comes to a discussion of traditional armaments, the U.S. Navy is a perfect example of how overbuilt our military is. According to Robert O. Work, senior defense analyst at the Center for Strategic and Budgetary Assessments, our current surface combat fleet has as much firepower as the next 20 largest navies (many of which belong to our allies) combined, and that is without considering carrier air wings or nuclear capabilities.[78] That means large savings can easily be had by eliminating unnecessary naval procurements.

The Sustainable Defense Task Force (SDTF) lays out an option that would save an estimated $126.6 billion over the next 10 years.[79] Their suggestions would cut the number of ships in the Navy's fleet from a planned 315 ships (by 2020) down to 230. These cuts would still leave the United States with overwhelming naval power for the next decade and beyond, so more could actually be done in the name of saving precious resources while still providing the Navy strategic options for its future fighting force. For instance, the SDTF suggests reducing the number of aircraft carriers from a planned twelve (in 2020) to nine, and reducing the number of naval air wings to eight. At present, the Navy is building its newest aircraft carrier, CVN 78. It also plans to procure CVN 79 and CVN 80, while retiring the USS Enterprise. But the United States could safely scrap the building of CVNs 79 and 80, and additionally retire the USS Nimitz and USS Eisenhower, reducing our carrier fleet to eight in 2020. An additional naval air wing could also be cut, reducing their number to seven. Reducing the additional carrier and air wing would save billions more than the SDTF has tallied.

The SDTF option for naval reductions mentioned above even leaves intact the Navy's procurement of the new Littoral Combat Ship (although at a lower number than the Navy currently has planned). This weapon system provides the Navy with a ship that can handle various mission packages — patrolling for drugs, terrorism or piracy, supporting humanitarian assistance and disaster relief efforts, helping to enforce sanctions, or securing access to coastal waters off foreign shores — essentially making it the Swiss

Army Knife of the future Navy. It also gives the Navy the opportunity to move away from a carrier-based fleet structure to a more distributed, networked battle force, if future entanglements so dictate. Even without cutting the procurement of this new ship, it's not hard to find $130 billion worth of savings over the next decade in naval expenditures alone.

Additional procurement savings could be realized by canceling the V-22 Osprey and F-35 Joint Strike Fighter. In both cases alternatives would have to be fielded.

For the V-22, ready alternatives are available in MH-60S and CH-53K helicopters. The Osprey is way over budget and its advantage over traditional helicopters is questionable. While the Osprey can fly faster than helicopters, making it less vulnerable to hostile fire when it's "on the move," it is much less stable than a helicopter in "hover mode," putting its crew and passengers at greater risk during drop-off and pick-up, particularly in hostile environments. The V-22s deployed in Iraq achieved a lower readiness rating than older helicopters in theater, even though the Ospreys had substantial manufacturer support. They also have trouble flying above 8,000 feet, operating in extreme heat, carrying an adequate number of troops, taking off from and landing on Navy ships, and carrying external cargo.[80] The fact that American taxpayers have already purchased 245 of these aircraft is more of a testament to industry lobbyists than it is to the operational capabilities of the vehicle itself. For the safety of our troops and the austerity of our budget it only makes sense to save up to $12 billion dollars over the next decade by canceling the Osprey and fielding alternatives.[81]

To replace the Joint Strike Fighter, the military could buy advanced versions of the F-16 and F/A-18E/Fs, depending on the branch of the service in need. This option was detailed in a CBO report in August 2009.[82] Updating the savings estimate to reflect current costs produces a budget reduction of roughly $58 billion for the next ten years.[83] Like many of the Pentagon's latest weapons developments, the F-35's cost trend is beginning to get out of hand. At some point a weapon system becomes so expensive that its cost outweighs the benefit it provides to the military, particularly when very capable alternatives exist. This is the case with the Joint Strike Fighter (JSF). All of the alternatives listed above are fourth generation aircraft or better, and give us a huge air advantage over the fighter wings of potential adversaries, most of which are composed of first- and second-generation aircraft. Even for countries like China that do have a number of third- and fourth-generation aircraft, their numbers of those aircraft are a fraction of ours. Moreover, schedule delays for the F-35 could cause serious problems because planned production rates for the JSF are insufficient to meet inventory goals as older aircraft are retired. Just as with the Osprey, canceling procurement of the Joint Strike Fighter actually allows us provide a *better defense* by spending *less money*.

Besides procuring less expensive aircraft, the Air Force could easily retire two fighter wings and create additional savings. As the Sustainable Defense Task Force points out, America's current air fleet has a much greater air interdiction capability than it did in previous years and our traditional adversaries haven't kept pace. Advocates for a larger military can theorize about the potential size of a future enemy's air force, but the reality is that the USAF is currently dominant by a wide margin and could afford fighter reductions without sacrificing security. Reducing two tactical fighter wings from the Air Force fleet would save approximately $40 billion.[84]

The Air Force could also delay procurement of the KC-X Aerial Refueling Tanker by five years, thus saving $10 billion dollars.[85] This would be done by upgrading 60 KC-135Es to the KC-135R standard. This delay would not only save money in the near future, it would allow the tanker procurement program to focus on newer commercial aircraft — the Boeing 787 or the Airbus A-350XWB. Both will most likely be cheaper to operate.

Approximately $11 billion could be saved over the next decade by canceling the Marine's Expeditionary Fighting Vehicle.[86] The SDTF suggests covering the Marines' desire for 573 of these vehicles by fielding refurbished AAV7A1s (currently used by the Corps as an armored amphibious vehicle) and an updated version of that model. But the last time the Marines made an amphibious assault was in 1950, over half a century ago, so the desire for 573 of these vehicles seems excessive, to say the least. A more reasonable plan would be for the Marines to refurbish the vehicles it currently has, but not procure replacements.

During the height of the Cold War, when America was engaged in a technology race with the Soviet Union, we spent $60 billion a year on research and development (R&D, adjusted for 2011 dollars). Today we spend almost $80 billion a year, although we have no superpower adversary to challenge us (some of this spending — roughly $7 billion — is due to the current wars in Iraq and Afghanistan).[87] The United States should trim these expenditures by $8 billion per year until the war in Afghanistan ends, and then increase the cuts to $15 billion per year past that point. There is no reason American taxpayers should have to pay for more R&D today (in real dollar terms) than they did during the Cold War. Given the comparative threats of the two eras, it doesn't make sense. This adjustment would save *at least* $80 billion over the next decade; a more realistic estimate would be on the order of $130 billion.

The Tenth Quadrennial Review of Military Compensation made numerous suggestions regarding the compensation and benefits that are received by military personnel and their families. Among these, it was suggested that the Department of Defense factor various elements — such as housing and subsistence allowance and tax advantages — into its pay raise calculations. It was also suggested that premiums for TRICARE, the military's healthcare system, be raised for retired service members under the age of 65. Although healthcare costs have gone up considerably, TRICARE premiums have not increased (for those under 65) since 1996.[*] Each of these suggestions would be phased in over a number of years, but taken together they would save the United States roughly $115 billion over the next decade.[88]

The 2009 Congressional Budget Office report, "Budget Options, Volume 2," lays out some suggestions for improving the efficiencies of military exchanges, commissaries and depots. Consolidating the four retail systems operating on military bases, changing the pricing structure at military depots, and allowing 10 percent more of depot work to be bid on by private contractors could save the Department of Defense $13 billion dollars during the period 2012 to 2021.[89]

Finally, with all the reductions mentioned above, it only makes sense that there would be less need for spending on

---

[*] Enrollees in TRICARE Prime are still paying the same premium they did in 1996, enrollees in TRICARE Standard and TRICARE Extra pay no premiums. TRICARE usually accounts for more than 8 percent of annual Defense Department spending.

headquarters, central support and infrastructure. The Sustainable Defense Task Force suggests that "a reasonable minimum goal for additional economizing would be 2 percent of the peacetime budget or approximately $10 billion per year," thus producing a final savings of $100 billion over the next 10 years.[90]

* * *

If you've been adding up the defense cuts mentioned above, you realize that roughly a trillion dollars of savings are possible over the next decade from the defense budget alone. More specifically, $980 billion to $1.08 trillion could be trimmed from our defense spending without really hurting our country's ability to defend itself.

But that's not to say that our country is currently well defended. The attacks of September 11th and the inability to bring a definitive halt to illegal immigration highlight serious weaknesses in our security posture. And while these have been addressed to some degree over the last decade with changes in intelligence gathering, transportation security, and tighter border control, more could — and should — be done. In addition to these cracks in our defense system, the threat of cyberwarfare and the problems it could create in the United States become more evident every day. The September 2010 discovery of a computer virus in Iran, developed to damage supervisory control and data acquisition systems at power plants and factories, shows how vulnerable a country's industrial and governmental computer systems can be in the 21st century. The Stuxnet virus, as it was called, ended up on tens of thousands of computers in Iran, Indonesia, and India. With the high degree of automation and computerization in America's defense and industrial systems, a worm like Stuxnet could wreak widespread havoc in the United States.

Therefore, over the next decade, $110 billion of the savings from these cuts to Pentagon spending should be directed back into programs that will actually help protect our country in the 21st century. The first focus should be a boosting of security patrols along the border. An additional 5,000 agents (above those promised by the current administration) should be added to the force that monitors not only our southern border with Mexico, but our northern border with Canada and our coastal waters along the Gulf of Mexico as well. These positions would serve to stop illegal immigration as well as halt narcotics traffic, and could easily be filled by retraining service men and women being released from duty in the armed forces. The second area of increased attention should be in shoring up America's cyber defenses, not only on the federal level, but the state and regional level as well. This spending would bolster the demand for well paying jobs in computer programming and network security, and prepare America for the threats of the 21st century. Finally, additional funds should be made available for better training and equipping first responders on the state and local levels. This would ensure that if a domestic attack does occur again, those first on the scene would have the skills and tools needed to minimize the damage.

It's important to note that these suggested changes in our defense spending not only save us hundreds of billions of dollars over the next decade; they actually make us safer than we are today. All one need do is look at the focus of the trillions of dollars the United States spent on defense in the decade leading up to 9/11 to see that we need a most desperate change of course. Because these changes focus on the most likely present and future threats — instead of spending money on "Cold War" weaponry prompted by some dreamed-up, unrealistic future enemy — our taxpayer dollars will be used more efficiently and effectively. Having the most expensive weapons systems isn't a constitutional requirement; providing a common defense is. These changes would ensure that we meet that responsibility.

One More Cut (At Least)

I pointed out earlier that balancing the budget would require a tough look at the government's annual tax collections, our entitlement programs of Medicare and Medicaid, as well as our outlays for the Department of Defense. And indeed, the tax changes and spending cuts mentioned up to this point will quickly get us to a balanced budget and even surpluses in the coming years (details in the next section). But that doesn't mean there aren't other areas of government spending that could be trimmed. While detailing a full budget breakdown would require a book unto itself, there is one area of spending that deserves to be mentioned because of the negative effects it has on our country and the large savings that could be had by reducing these expenditures. That portion of the budget is the farm bill.

The origins of our government's farm aid programs date back to the Dust Bowl and the Great Depression of the 1930s. What once was a program intent on saving small family farms now subsidizes much larger farms that don't need the money. Roughly 75 percent of U.S. farm aid goes to the richest 10 percent of farmers. And although our agricultural sector accounts for only about 2 percent of American jobs, it takes about one-half of the direct and indirect subsidies coming out of Washington.

The farm bill is a perfect example of why the government shouldn't address economic problems with welfare programs or business subsidies. Remember, the main problem during the Great Depression was an extreme level of income disparity — the same problem we have today. Farm subsidies, or any business subsidies or welfare programs for that matter, don't rectify that actual problem; they simply mask it with government expenditure. Once a subsidy is in place, it is simply too difficult to remove, even if the economic problem it was meant to fix wanes. That should be evident in the fact that our government continued to dole out farm subsidies in 2008, a year when commodity prices were through the roof.

Not only do our farm subsidies provide misguided economic help, there's strong evidence that these handouts actually hurt our country in numerous ways. First, subsidies help American farmers grow corn on large industrial farms in the United States that can be sold below cost in places like Mexico, where Mexican farmers simply can't match the price. Once the Mexican farmers are driven out of the market and can no longer make a living farming in their home country, they come north looking for work and become America's illegal immigration problem. The second problem caused by the farm bill has to do with our health. Corn subsidies drive down the cost of high-fructose corn syrup, and in turn, all the junk food made with that ingredient. Because junk food is cheaper than locally grown fruits and vegetables, almost never subsidized, it has become the "go-to" choice for America's families. Once again, these poor health habits drive up the demand and cost for health care, along with spending on Medicaid and Medicare. (Nothing like having our tax money wasted on a government program that eventually costs us more tax money in another government program!) Finally, these farm subsidies are always a sticking point in trade negotiations with other countries. Eliminating, or at least reducing the subsidies not only makes sense in regards to establishing a trade regime that's fair, but could help to cool any animosities potentially caused by America's new tariff regime.

Starting in 2012, the United States should begin to reduce its subsidies to farmers. The first year, subsidies should be reduced by 55 percent of what is currently budgeted; that percentage should rise to 65 percent in 2013, and then to 75 percent from 2014 through 2021. These cuts would produce a ten-year savings of just over $110 billion dollars (not to mention that they would eventually reduce some of our other problems caused by their existence).

Doing the Math: The Results So Far

So what exactly do all of these cuts produce? Do these savings produce a balanced budget today, or in the near future? Let's look at the numbers.

Balancing the budget immediately is simply too difficult (and unwise) a task, regardless of how good the plan is. And so, like all other deficit reduction proposals introduced to date, this plan calls for continued deficits in fiscal years (FY) 2012 through 2014. However, these projected deficits are smaller than those presented in other proposals, and, unlike other proposals, this plan would produce a slight budget surplus by FY2015.* These surpluses are projected to grow, so that in FY2021 the budget surplus would be over $250 billion.+ (It is important to note that I have not factored in savings from spending cuts that could not be reasonably quantified, so the budget figures expressed here are likely to be somewhat conservative.) Over the ten years from FY2012 to FY2021, this economic and fiscal plan would produce an overall surplus, reducing the national debt by $45 billion. In comparison, the budget, as it is currently projected, will run deficits of at least half a trillion dollars in each fiscal year 2012 through 2021, and will add almost $7 trillion dollars to the national debt.[91]

Fixing Social Security

While the improvement in the budget situation mentioned above is good news, it hides a fiscal problem that has yet to be

---

*This is a combined budget estimate, on-budget and off-budget included.

+ Once again, this is a combined figure. The on-budget surplus for that year is projected to be almost $210 billion.

addressed — the long-term outlook of the Social Security system. Although not in a crisis, as some politicians have claimed, the Social Security system does have a negative long-term outlook. The outlook is bad enough that, according to the CBO, the "resources dedicated to the program will become insufficient to pay full benefits in 2039."[92]

The way Americans view Social Security today is very different from the original intent of the program. Social Security was originally established, not as a government mandated retirement fund, but as a social insurance program. The system was set up to protect Americans against the unexpected difficulties inherent in American life — disability, death of a family's breadwinner, unemployment, and the inability to earn a living due to old age. The contributions that each worker made to the system were not viewed as a tax, but as a sort of insurance premium — hence the name Federal Insurance Contribution Act (F.I.C.A.), regularly taken out of your paycheck. The premium, collected by the government, contributed to a fund that insured against unfortunate occurrences that might befall the worker and impair his or her ability to provide for his/her family. If a worker became disabled, the government "paid out on the policy," so to speak. If workers became so advanced in age that they could no longer work for a living, the government provided them a means to live.

In time, Social Security stopped paying out benefits to those who had become unemployed, but it did still provide payments to the disabled and enrollees who became too old to work. To that end, the original eligibility age was set high enough that a lot of workers never attained it. When the system was first created, the eligibility age was set at 65. Since the average life expectancy at the time was just under 62 years, a lot of people never gained access to the benefits, and those who did typically didn't collect benefits for very long.

Although the Social Security Administration claims that the life expectancy figure at the time was skewed by high infant mortality rates, even their figures show that of those Americans who reached adulthood (age 21), less than 54 percent of men and just over 60 percent of women ever made it to age 65.[93] In other words, it was expected that a lot of people would work until they died. (Of course, if you expected that of people today, they'd think you were nuts!)

As time passed, life expectancy increased. But the government never adjusted the eligibility age for retirement benefits (until recently) so more people reached the age where they were eligible for the benefits. In addition, those beneficiaries typically drew upon the benefits for a longer period of time. This caused two problems.

The first is that it transformed people's view of the program — from one of a social insurance system that provides a safety net against the inability to work in old age, to one of a retirement program run by the government. Part of the reason that the Social Security system is now so difficult to reform is that Americans have come to view the program as a retirement fund. This viewpoint is so prevalent that politicians and pundits (who should be more informed about the program before they speak publicly) make references regarding the "return" on the program, as if it were an investment fund. Or maybe they speak about proposed changes to the system in terms of what is "fair" or "unfair" to a given group of participants in the program. This is really the wrong way to evaluate the Social Security system. Numerous Americans pay for car insurance, but never collect anything from their insurer, because they never have an accident. No one views that as "unfair," or a "poor return on the investment" because they realize that it's insurance. This is ultimately what Social Security was supposed to be — social insurance.

The second problem is that it has caused long-term budgetary issues within the Social Security system. With more people becoming eligible to receive benefits, and collecting them for a longer period of time, the system developed long-term solvency issues.[*] As these financial problems became apparent, the government began raising the premiums that workers and employers paid into the system. When initially created, the system charged a fee of 2 percent (1 percent from the employee, 1 percent from the employer) of eligible earnings; today that rate is 12.4 percent. This, of course, reinforces the view of Social Security as a retirement program rather than a social insurance program; 6.2 percent of one's

---

[*] It's also important to note that numerous changes have been made over the life of the program, some of which expanded coverage or increased benefits. While these have also negatively affected the long-term balance of the system, I'll confine myself to societal changes that have had the greatest impact on the system.

paycheck (matched by the employer, for a total of 12.4 percent) seems more like a 401K contribution than an insurance premium.

While raising the premium might have been necessary at certain times, the real adjustment the government should have made was to increase the eligibility age of enrollees to match increases in the average American lifespan. If the average life expectancy today were 120 years, would it really make sense to have a retirement age in the mid-sixties, so that people work only half their lives? Of course not. The government finally took the right step in 1983 by increasing the eventual eligibility age from 65 to 67, so that those born after 1937 face a higher eligibility age. According to current law, anyone born after 1959 isn't eligible for full Social Security benefits until age 67. But the eligibility age should actually be raised some more. Today life expectancy in the United States is over 78 years of age and eligibility requirements are still 67 years of age or less.

So the first adjustment that should be made is to once again increase the eligibility age for people to draw from Social Security. The chart on the next page outlines a suggested change in the eligibility age of Social Security in comparison to current law. Gradually lifting the eligibility age to 70, as shown in the chart, would roughly match the advancement in life expectancy that has taken place since 1940. This option has been analyzed by the Congressional Budget Office. The CBO estimates that it would eliminate roughly half of the projected $5.3 trillion shortfall in the Social Security System over the next 75 years.[94]

Our country has to get back to the realization that the Social Security program is a safety net, not some sort of social hammock that allows people to retire in their sixties, regardless of how old they grow to be. Therefore, once the full retirement age reaches 70 for those born in 1978, it should be indexed to changes in longevity. This would keep the Social Security system from growing more beneficial (and costly) as Americans live to grow increasingly older. Note that these modifications don't prohibit Americans from retiring early. Workers can still retire at age 65, or earlier, if they choose. They just have to do so on their own dime — without government assistance. These changes would help counter the view that Social Security is a retirement program and replace it with the program's original role as an old-age assistance program. They might also provide an incentive for younger workers to save for their own retirement. And with the economic changes introduced earlier, those younger workers would actually have the means to do so.

Social Security Eligibility Age

Birth Year	Current Eligibility Age	New Eligibility Age	# of additional months a retiree must work:
1959	66 and 10 months	66 and 10 months	0 more months
1960	67	67	0 more months
1961	67	67 and 2 months	2 more months
1962	67	67 and 4 months	4 more months
1963	67	67 and 6 months	6 more months
1964	67	67 and 8 months	8 more months
1965	67	67 and 10 months	10 more months
1966	67	68	12 more months
1967	67	68 and 2 months	14 more months
1968	67	68 and 4 months	16 more months
1969	67	68 and 6 months	18 more months
1970	67	68 and 8 months	20 more months
1971	67	68 and 10 months	22 more months
1972	67	69	24 more months
1973	67	69 and 2 months	26 more months
1974	67	69 and 4 months	28 more months
1975	67	69 and 6 months	30 more months
1976	67	69 and 8 months	32 more months
1977	67	69 and 10 months	34 more months
1978	67	70	36 more months
1979	67	70	36 more months
1980	67	70	36 more months

Of course, solving half the problem isn't enough. We need to make other adjustments to the Social Security system in order to bring about complete long-term solvency.

Since Social Security was originally established, in part, as an insurance against the inability to earn a living in advanced age, it only makes sense that those who are earning adequate incomes, despite their age, shouldn't need to collect on the policy, or at least not collect as much. The government recognized this to some degree when it made a portion of a retiree's benefits, above a certain threshold, subject to the federal income tax. This change, instituted in 1984 under the Reagan-era Social Security reforms, made 50 percent of an enrollee's benefits subject to the income tax if that person's income was over $25,000 a year (for a single filer). The portion that qualified for taxation was then raised to 85 percent under the Deficit Reduction Act of 1993.

Since this economic plan completely eliminates income taxes for the vast majority of Americans, these former percentages of a person's benefits that are subject to the income tax no longer make much sense. So to reduce the benefits taken by those retirees who don't really need them, we should *gradually* lower the initial benefits for the top 50 percent of beneficiaries. This can easily be done, without cutting benefits for current recipients, through what is known as "progressive price indexing."

Under the current system, a beneficiary's primary insurance amount (PIA) is determined by how much that person earned in the highest 35 years of earnings on which he or she paid F.I.C.A. Dividing the sum of those earnings by 420 (35 years times 12 months per year) produces what is known as a person's "average indexed monthly earnings" (AIME).* A worker's AIME is then multiplied by a PIA formula to determine the monthly Social Security benefit. The formula is set up so that a greater percentage of income is replaced for lower wage workers. In 2010, the primary insurance amount formula was 90 percent of the first $761 of average indexed monthly earnings, plus 32 percent of a person's AIME between $761 and $4,586, plus 15 percent of AIME above $4,586. The factors of $761 and $4,586 are known as "bend points," because they are points at which the benefit curve changes and replaces less of a person's AIME. The graph on the following page should help to demonstrate.

Each year these bend points are adjusted for changes in the average annual earnings of the country's workforce.

Because the Social Security system is meant to be a safety net, and not a retirement fund, the second adjustment that we make should be to slowly reduce benefits at the upper end of the scale until there is eventually a cap in place that limits the amount of monthly benefits an enrollee can receive. After all, those who have had higher

---

* A person's annual earnings from younger years are also adjusted for inflation and real wage growth.

earnings throughout their life should need fewer Social Security benefits, not more. And if they do fall on hard times, for whatever reason, the amount that Social Security *does* provide should be enough to keep them out of poverty. After all, that's what the purpose of a safety net is, right?

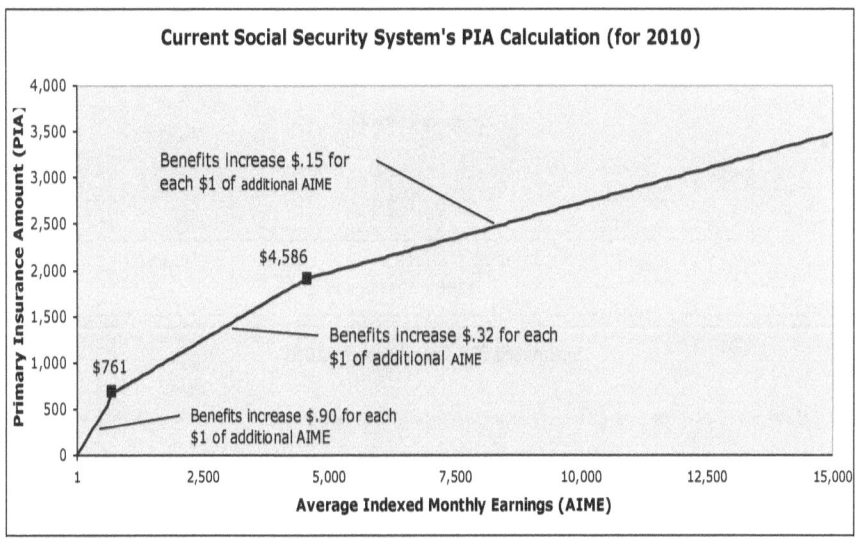

The way that this adjustment would be accomplished would be to add a third bend point between the first bend point and the second (see charts below). Benefits beyond that point would still be adjusted for inflation, but not completely increased for changes in average real earnings. For newly eligible beneficiaries who earned the taxable maximum during the 35 years used in the AIME calculation — maximum earners — benefits would be adjusted for inflation, but not at all for changes in the real earnings of society. For those who fall between the new bend point and the maximum earners, benefits would be modified to account for inflation and part of the change in real earning, but not the entire amount. This would slowly reduce the increase in benefits received each year by the top 50 percent of wage earners. The following two charts give a graphic representation of how benefits would change over time. Notice that by 2051 Social Security benefits will be capped at an estimated $2,725 per month (this amount would be adjusted each year for inflation) so that no one, regardless of income level during their working lives, would receive more than that.

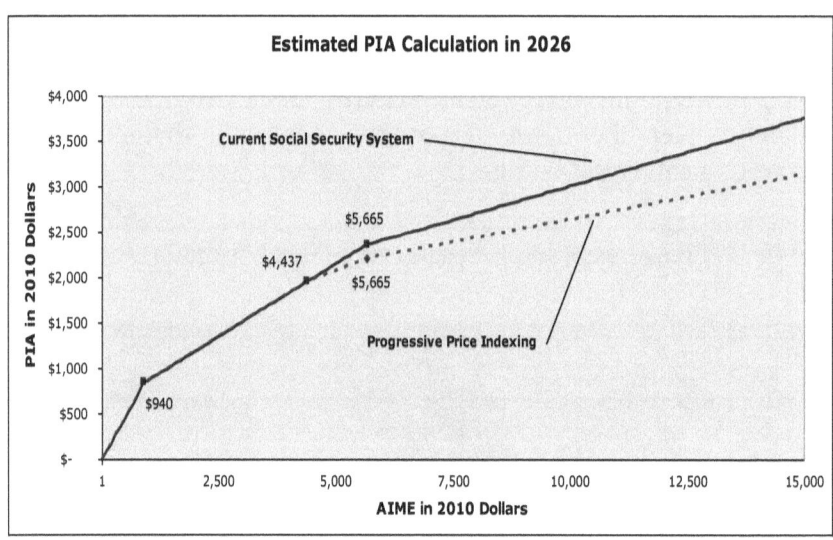

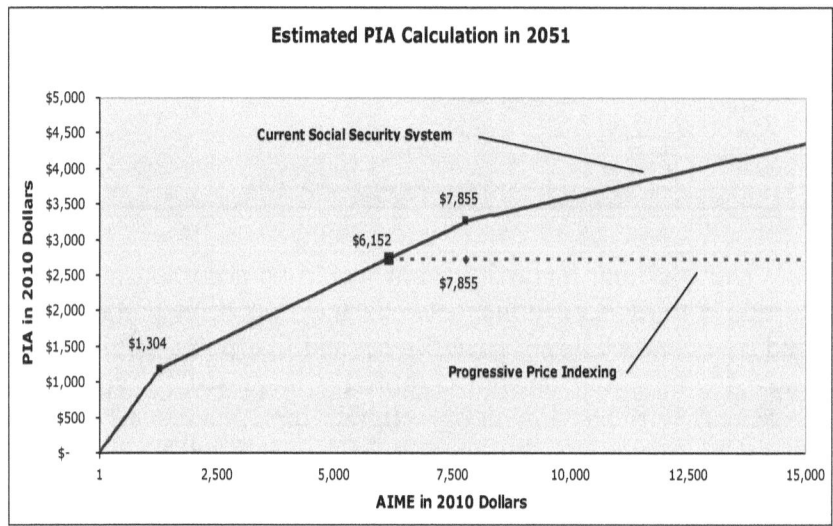

According to a July 2010 report by the Congressional Budget Office, the two changes mentioned above would be more than adequate to cover the 75 year projected shortfall in the Social Security System.[95] In fact, they would even allow some room to make upward adjustments in Social Security payments for poorer beneficiaries to help cover the cost of the new BTU tax. These modifications gradually return Social Security to the sustainable safety net program it was intended to be, while protecting current recipients who need it most.

Conclusion

As with Medicaid and Medicare, this plan makes Social Security more sustainable, not simply by reducing future expenditures, but by giving those who are most likely to rely on these programs a better opportunity to provide these needs for themselves. By raising wages for the majority of workers, Americans would have the chance to provide for themselves what many of them are having a hard time providing: health care in the present and healthcare and retirement savings for the future. Thus we will not only have a feasible plan for bringing future budgets into balance, but one that is fair to less fortunate Americans. Combined with additional revenue, and immediate savings from defense spending and agricultural subsidies, this plan could easily produce a balanced budget (and even a slight surplus) over the next 10 years.

These savings will pay additional dividends as well. By balancing the budget by FY2015, and beginning to run surpluses, the United States will save roughly $1.5 trillion in interest expenses from FY2013 to 2022. That's $1.5 trillion that won't have to be produced by raising taxes or cutting additional services.

Conclusion

As this book goes to press, the economy seems to be slowly slipping back into a potential recession. The two years of "official" recovery we've had have not only been brief, but they've done little to help most average Americans. With unemployment barely dipping below 9 percent, wages seeing no real growth, at least for most workers, and the housing market still staggering around like a fighter being administered an eight-count, the fact that there has even been a recovery is probably a wonder to most people.

Sadly, Washington has once again failed to learn the lessons being offered up by the country's pain; both sides are offering up the same old failed prescriptions for bringing the country's economy back to life. Democrats, for the most part, have suggested more government spending (labeled now as "investment"), despite the fact that the government is now spending more than it ever has, with few results. Republicans have countered, as usual, with the gospel of tax cuts, particularly for corporations and those in America's top income bracket. With taxes currently near a low point across the board, for corporations and individuals, and the top 400 income earners paying an average of only 17 percent,[96] one would think the economy would be booming, but it's not.

Let us not forget that the Federal Reserve has been in the battle as well, printing enough money to smother the nation in paper. The fact is that this latest recession has been battled with all three traditional methods of economic stimulus — tax cuts, spending increases, *and* a generous creation of additional money from the Fed — and all that these steps have produced are a huge increase in government debt and a recovery nobody seemed to notice. The obvious lesson that these traditional remedies won't work must have gotten lost in all the partisan bickering between the left and right of the political spectrum.

We seem to have forgotten the lesson of Henry Ford, who insisted on paying his workers twice the going rate for their labor. Ford realized that if workers didn't make enough to afford the cars his company was producing, his company would fail. The economy is no different, and that principle is just as true today as it was in Ford's day. Somewhere in the last 40 years we have forgotten that. Rather than making sure workers had jobs and those jobs paid living wages, we allowed someone to convince us that we could hide our income problem with more government transfer payments. Or maybe we were convinced that if we let those at the top take as much as they want, and give them a tax cut to boot, the money would eventually trickle down to all of us and make everything good again. Maybe we believed that we could import cheaper products from low wage countries, even if it cost American jobs, and the economy would continue magically growing as normal. Or maybe we just thought we could finance it all with a credit card or an additional home equity line of credit, and someday we'd miraculously be able to pay it all back.

Whatever the case, the game is up; our economy has flat-lined and our national bank account is empty (to say the least). Instead of simply making sure the economy was fair for all participants and then putting the responsibility on individuals for taking care of themselves, the government has permitted the economy to favor an extreme few and then taken on the responsibility for taking care of the many in order to rectify the imbalance. The continual tax cuts, food stamps, government-backed student loans, extended unemployment benefits, jobs programs, earned income tax credits, retraining programs, business tax credits, expansion of safety-net programs, corporate bailouts, and all the other gimmicks our government has tried over the years have not been enough to make up for the fact that most American wages have simply fallen short. The better part of the American workforce no longer has the income necessary to supply the demand needed to make our economy prosper. And because the overabundance of savings — produced by the fortunate few who have incomes well beyond what they want and need — is greater than the supply of productive investments, we experience investment bubble after investment bubble, further endangering our already fragile economy.

If it isn't obvious at this point, let me state it one more time: *The only way to fix our problems is to first fix our excessive level of income disparity!*

The goal of this plan is to do precisely that. And while we could probably develop numerous ways to rectify the income disparity problem, it would be best if we did it in a way that allowed us to tackle the other secondary problems of our nation as well. Too many solutions that come out of Washington are focused on solving one problem, and therefore end up causing harmful side effects in their wake. This plan is a comprehensive plan — one that solves our economic problem, but does so in a way that also allows us to balance the budget, reduce our energy usage, and lessen our impact on the environment.

The Plan

The first step in rejuvenating our economy should be to produce an income scale that is compact enough to keep the economy functioning properly without continual government assistance. This should be done by raising wages on the lower end of the pay scale and reducing effective income at the top. The best way to begin that process is to gradually increase the minimum wage until it reaches a meaningful level. This will put a floor under wages that should prevent workers from needing to seek assistance from the federal government. It will also put upward pressure on the entire bottom half of the income spectrum, improving the lives of tens of millions of workers. Halting illegal immigration, supporting legitimate demands of labor unions, and enforcing antitrust regulations will assist this wage increasing process by shifting economic power from corporations to workers.

Because the wage floor will be increased slowly, but our economy could use a greater economic stimulus in the near term, we should reinforce the stimulus effect with income tax changes. Eliminating income taxes for the majority of workers will not only provide a strong boost to the economy, but makes sense from an economic perspective in that taxing labor reduces the demand for it. This system also greatly simplifies the overall income tax code. And while eliminating the lower income tax brackets will provide an income tax cut to almost everyone making up to $500,000 a year, some of the largest tax breaks will be received by those in the middle and upper-middle class (those earning $90,000 to $150,000 a year, depending on filing status).

Of course reducing income taxes for over 99 percent of Americans will reduce government revenue. This will be made up by increasing taxes on outrageously large incomes. And while this step will no doubt be opposed by conservatives and business elites, it actually makes good sense. We saw in Chapter 2 that objectively justifying these incomes is difficult and that they stem more from greed and imperfections in the market system than from any sort of effort or productivity. Rewarding executives with millions of dollars, even for poor performance, can't be explained with any reasonable economic thought. Neither can the rewarding of small differences in performance with millions of dollars of difference in income. The lack of restraint on these types of compensation over the last three decades has resulted in extremely large income gains for the top half of one percent of income earners, while the real income of common workers has been stagnant at best. In other words, the income gains that should have been shared by all of us — after all, we were all responsible for the country's productivity gains — were stolen by a small minority of workers. Having a high top marginal tax rate that only affects the highest of incomes will deter this greed and reduce the incentive for these few individuals to take advantage of their positions in order to unduly increase their income at the expense of everyone else.

The effect of these income increases and tax changes will be an increase in economic activity and a boost for small businesses. With ordinary workers making more money, they will be able to patronize small businesses instead of being forced to shop at "big box" stores. Releasing most workers from the burden of paying income taxes will also reduce the burden on businesses for regularly sending in these tax payments, making businesses easier to manage financially. The new income tax system, with two simple rates that won't affect most small business owners, will give small business owners the opportunity to plow profits back into their venture, rather than having them taxed away by the government, giving them a greater likelihood of success. And the changes to capital gains taxes will ensure that there is a ready pool of investment funds waiting for new ventures if a small business owner decides that she needs more capital to speed expansion.

The economic effects will be coupled with important societal changes as well. With workers finally seeing real income gains, the daily stress brought on by financial uncertainty will slowly decrease. There will be less pressure on families to send both parents into the workforce. The follow-on effects could easily be stronger family ties, healthier diets, less crime and drug addiction, and fewer abortions and divorces. College graduates might actually have the hope of getting a job that allows them to move out of their parents' homes after graduation.

But this system won't work if we continue to allow imports from low-wage countries destroy the American industrial base. We must therefore enact a tariff schedule that recognizes and allows the benefits of a free-trade system, but also levels the playing field for American workers on the international stage. In other words, if foreign companies can produce a product more efficiently than American companies, then American consumers should be allowed to enjoy those savings by importing the product with little or no tariff. Asking American workers to compete against companies whose workers are paid a lot less, sometimes a tenth of American wages, is an impossible request. The tariff system outlined in Chapter 5 would simplify the tariff schedule of the United States, allow consumers to enjoy the proper benefits of free trade, *and* produce a fairer playing field for American workers. Coupled with the income and tax changes mentioned earlier, this new tariff system should spark a strong economic recovery as retailers begin to look for domestic suppliers rather than relying on cheap imports for their inventory.

This economic plan isn't just a short-term recovery plan, though. It lays the foundation for long-term economic growth and a resolution of many of our country's other problems.

While the increases in the minimum wage and the changes to the income tax system will help to produce a distribution of the country's income that is fairer, it is obviously *not* fair to have a small percentage of Americans paying all the taxes (roughly 5 percent would be paying income taxes). Therefore this economic plan would introduce a BTU (British Thermal Unit) tax that would be paid by all Americans. The amount of the BTU tax starts small, but gradually increases over the first decade of implementation. This is done for a few reasons.

The first is that it allows the economic expansion to take hold, rather than squashing it with the fully implemented amount of the BTU tax. The second is that poorer Americans, who aren't presently paying federal taxes, will be paying this tax; increasing it gradually, as their incomes increase, will allow them to contribute without unduly burdening them. (The yearly increase in the BTU tax is calculated so that even someone working full time at the minimum wage would be able to afford this tax and still see an increase in take-home pay.) Finally, the gradual implementation will put America on the path to saving our energy resources without overly disrupting the current fossil fuel system that now supports our economy to a large measure. Moreover, it will get us moving toward developing cleaner, renewable energy alternatives without requiring the large government investment that many Democrats suggest.

Because this plan would revive the economy without massive spending increases or large losses of government revenue; because it would reduce our energy use without government investment; because it would lessen our environmental impact without new regulations or government bureaucracy; because it would establish an economy where workers can pay their own way rather than relying on government programs, it also establishes a foundation for reducing the size and scope of government, balancing the budget, and beginning to pay down the national debt. And although the changes to the income tax system, the changes to our tariff schedule, and the creation of a BTU tax will bring in additional revenue over the next decade, expenditures will still have to be reduced in order to balance the budget and begin running a surplus.

The largest cuts to federal spending in this plan will come from Medicare, Medicaid, and the Defense Department. These budget busters are all long overdue for reform, and the road map put forth in Chapter 7 detailed ways that each can be cut. With over a trillion dollars cut from Medicare and Medicaid, roughly a trillion dollars saved from cuts to military spending, and additional savings coming from a reduction in agricultural subsidies, the government could be back to producing yearly budget surpluses before the 2016 presidential election.

So Why Not? A Clarion Call to Arms

Despite the rational basis behind this plan and its comprehensive nature, it will not come easy, nor be free. There will likely be opposition to various parts of this proposal from either or both sides of the political spectrum. Americans must be willing to let go of the myths that have been perpetuated by our two dominant political parties and be willing to vote in ways that are different from our normal selection processes, considering third-party candidates or even independents. We must think through the explanations that our leaders give us for enacting various policies and make sure that the logic behind those decisions makes sense. For many people this will not be easy.

Americans have a strong belief in the tenets of free market capitalism. And while it is the best economic system the world has known, it is not perfect. We must recognize the flaws in this system and be willing to correct them in a way that does not disrupt the general function of the market, or create an oversized government. Some politicians and pundits will argue that the increase in the minimum wage and the high top marginal tax bracket are impediments to the natural functioning of the market. But these aspects of the plan simply replicate the conditions that we would see under the assumed conditions of the purely competitive market — higher wages due to full employment and lower returns to capital owners produced by greater competition. And while conservatives will no doubt argue that this highest tax rate will reduce incentives to work and produce, we must remember that this is only a problem in economies operating at full capacity (ours certainly is not) and that incomes at these extreme levels have little to do with effort anyway.

There will be allegations from those on the political right that this plan institutes nothing more than class warfare. While this argument might be valid, we must realize that class warfare takes place every day in this country. When an executive back dates his stock options, costing the corporation money that could go to workers or shareholders, that is class warfare. Any time a CEO receives an undue pay raise from a corporate board packed with friends and allies, this leaves less for workers and causes an increase in prices for consumers. That is class warfare. And when wealthy Americans influence our government officials to benefit themselves at the expense of other Americans, that too is class warfare. Class warfare has allowed the top half of one percent of income earners to steal most of the income gains we've seen over the last three decades. What would be wrong with using class warfare to make sure those gains go to all workers? Americans must ask themselves why Republicans complain so vociferously about some types of class warfare, and not about others.

Democrats will no doubt complain that the suggested cuts to Medicare, Medicaid, and Social Security will destroy our country's safety net and leave the elderly and less fortunate without income. The argument is false. The plan described in this book only makes adjustments to Medicare and Social Security payments for future beneficiaries and those who could easily afford the reduction in benefits. In fact, these changes would strengthen both programs by making them more sustainable for the long term. And while the cuts to Medicaid may seem dramatic, this plan will improve wages and employment enough that the program will be much less necessary. In fact, the income improvements engendered by this plan should make the whole country less reliant on our country's safety net. With reasonable solutions such as this, one must wonder why the Democrats are so determined to have Americans so reliant on the federal government.

There will most likely be opposition from both major political parties to the idea of rejecting universal free trade with all nations. This is despite the fact that for hundreds of years it has been recognized that with unrestricted trade in place, employers will move work from high wage countries to countries with lower wages. Opponents will probably argue that the tariff plan presented in this book will provoke China to sell our government's debt on the open market, driving up interest rates and wrecking our economy. While such a response from China is certainly a possibility, the effect will be muted by the fact that this plan balances the federal budget so quickly and begins to pay down the national debt with yearly surpluses.

These arguments, presented in the previous few paragraphs, make little or no sense when we consider the full picture of how our economy operates, or the potential solutions available to our nation. And yet they are typical of the flimsy responses we get from both parties in regard to solutions — both liberal and conservative — that could actually work. In the place of workable solutions, we get variations of the same old policies — from each side — that have continually failed us in the past and frozen Washington in ideological gridlock.

The plan presented in the previous pages is an attempt to break that logjam with a set of proposals that, in some ways, should appeal to both partisan sides. But in order to put it in place and gain its benefits, the American people must show that they are willing to rip the power away from our two dominant political parties and give it to leaders who see the simple logic detailed in this type of economic plan. Unless the voters of this country rise up and show that they are willing to vote for their own economic benefit — instead of constantly supporting someone else's — the Democrats and Republicans will continue to lead us toward greater income disparity, a larger government, and more government debt. This is a losing proposition, one that only "we the people" can change.

References

Chapter 1 – America's Problems
[1] Emmanuel Saez, "Striking it Richer: The Evolution of Top Incomes in the United States," University of California, Department of Economics, August 5, 2009
[2] Thomas Piketty and Emmanuel Saez, "Income Inequality In The United States, 1913-1998," *The Quarterly Journal of Economics* Vol. CXVIII, February 2003
[3] Emmanuel Saez, "Striking it Richer: The Evolution of Top Incomes in the United States," University of California, Department of Economics, August 5, 2009
[4] Alvin Josephy, "The U.S.: A Strong and Stable Land," *Time*, Sept. 14, 1953
[5] Piketty and Saez, "Income Inequality in the United States: 1913-1998," *Quarterly Journal of Economics*, July 2008 revision
[6] Bureau of Labor Statistics; inflation adjustment calculation based upon CPI-U, provided by the Bureau of Labor Statistics
[7] U.S. Census Bureau, Current Population Survey, Annual Social and Economic Supplements, 2009
[8] Mishel, Lawrence, Jared Bernstein, and Heidi Shierholz, "The State of Working America 2008/2009," *An Economic Policy Institute Book*, Ithaca, NY; ILR Press, an imprint of Cornell University Press, 2009
[9] Jeanne Sahadi, "CEO Pay: 364 times more than workers," CNNMoney.com, August 29, 2007
http://money.cnn.com/2007/08/28/news/economy/ceo_pay_workers/index.htm
[10] Federal Reserve Board, "Federal Reserve Statistical Release – Consumer Credit," June 7, 2010
http://www.federalreserve.gov/releases/g19/current/
[11] Ravi Batra, *The Great American Deception: What Politicians Won't Tell You About Our Economy and Your Future,* John Wiley & Sons, Inc., 1996
[12] The National Marriage Project (University of Virginia), "New Report Finds Marriage Trouble in Middle America," Press Release, December 6, 2010

[13] The National Marriage Project (University of Virginia), "The State of Our Unions," The National Marriage Project and the Institute for American Values, December 2010

[14] Jones RK, Darroch JE and Henshaw SK, "Patterns in the socioeconomic characteristics of women obtaining abortions in 2000–2001," *Perspectives on Sexual and Reproductive Health*, 2002, 34(5):226–235

[15] Finer LB et al., "Reasons U.S. women have abortions: quantitative and qualitative perspectives," *Perspectives on Sexual and Reproductive Health*, 2005, 37(3):110–118

[16] http://vegetablegardens.suite101.com/article.cfm/reviving_the_victory_garden

[17] http://www.guardian.co.uk/science/2006/jan/15/socialcare.food

[18] Paul Krugman, *The Conscience of a Liberal*, W. W. Norton & Company Ltd., 2007

[19] U.S. Census Bureau, Current Population Survey http://www.census.gov/hhes/www/cpstables/032009/hhinc/new06_000.htm

Chapter 2 – Understanding Economics, Capitalism, and Why the Traditional Remedies Usually Fail

[20] Carolyn Lochhead, "Ranchers Accuse Meatpackers of Price-fixing", San Francisco Chronicle, Sunday, February 6, 2011

[21] Robert H. Frank and Philip J. Cook, *The Winner-Take-All Society: Why the Few at the Top Get So Much More Than the Rest of Us*, Penguin Books, 1996

[22] Malcolm Gladwell, *Outliers: The Story of Success*, Little, Brown and Company, 2008

[23] National Association of Corporate Directors, Webcast: "Trends in Director Compensation: Preliminary Findings from the 2009-10 NACD/PM&P Director Compensation Report" January 21, 2010

[24] Lucian Bebchuk and Yaniv Grinstein, "The Growth of Executive Pay," *Oxford Review of Economic Policy*, Vol. 21, No.2, 2005

[25] Erik Lie, "On the Timing of CEO Stock Option Awards," *Management Science* Vol. 51, No. 5, May 2005

Chapter 3 – The Road to Ruin

[26] Carrie Johnson, "Medical Fraud a Growing Pain," *The Washington Post*; June 13, 2008; http://www.washingtonpost.com/wp-dyn/content/article/2008/06/12/AR2008061203915.html

[27] Internal Revenue Service, "Earned Income Tax Credit (EITC) Questions and Answers," http://www.irs.gov/individuals/article/0,,id=96466,00.html

[28] Internal Revenue Service website: http://www.irs.gov/individuals/article/0,,id=177571,00.html

[29] Charts produced from information contained in the "2008 Economic Report of the President" 2008

Chapter 4 – A New Capitalism

[30] Possibly the best of these studies is the Card and Krueger analysis of effects caused by the minimum wage increase in New Jersey in 1992 (not only for the work of the authors, but the ideal conditions under which the study took place). Below, I cite both the original study as well as the reply that was necessitated by criticism of their work. More specifically – studies by Richard Berman of the Employment Policies Institute (EPI) and David Neumark and William Wascher both questioned the findings of Card and Krueger and suggested the opposite effect. A full analysis of the studies shows the EPI/Neumark and Wascher studies (both based, in part, on the same data) to be highly questionable and the work of Card and Krueger to be more reliable.

David Card and Alan B. Krueger, "Minimum Wages and Employment: A Case Study of the Fast-Food Industry in New Jersey and Pennsylvania," *The American Economic Review* Vol. 84 No. 4, September 1994

Reply: David Card and Alan B. Krueger, "Minimum Wages and Employment: A Case Study of the Fast-Food Industry in New Jersey and Pennsylvania: Reply," *The American Economic Review* Vol. 90 No. 5, December 2000

[31] Bureau of Labor Statistics, "Labor Force Statistics from the Current Population Survey – Characteristics of Minimum Wage Workers: 2009" **http://www.bls.gov/cps/minwage2009tbls.htm**

[32] Bureau of Labor Statistics, "Occupational Employment Statistics (Downloadable Occupational Employment and Wage Estimates) – May 2009 Occupational Employment and Wage Estimates," **http://www.bls.gov/oes/oes_dl.htm**

[33] Edward N. Wolff, "Recent Trends in Household Wealth in the United States: Rising Debt and the Middle-Class Squeeze – An Update to 2007," Working Paper No. 589, Levi Economics Institute of Bard College, March 2010

[34] David Cay Johnston, "Dozens of Rich Americans Join In Fight to Retain the Estate Tax," *New York Times*, February 13, 2001

[35] IBID

[36] *Franklin Roosevelt, in a speech to Congress* (Message to Congress on Tax Revision
June 19, 1935),
http://www.ustreas.gov/education/faq/taxes/historyroosevelt message.shtml#content

[37] Theodore Roosevelt, "State of the Union Address to Congress," December 3, 1906

[38] Citizens for Tax Justice, "Who Pays the Federal Estate Tax," April 4, 2006
http://www.ctj.org/pdf/est0406.pdf

[39] George J. Borjas, "The Labor Demand Curve *Is* Downward Sloping: Re-examining the Impact of Immigration on the Labor Market," *The Quarterly Journal of Economics*, November 2003

[40] United States Department of Agriculture, "2007 Census of Agriculture – United States Summary and State Data," February, 2009

Chapter 5 – A New Trade

[41] Hume (1748), quoted by Kym Anderson, *New Silk Roads: East Asia and World Textile Markets*, Cambridge University Press, 1992

[42] David Ricardo, *On the Principles of Political Economy and Taxation*. 1821, Library of Economics and Liberty, October 11, 2011,

http://www.econlib.org/library/Ricardo/ricP.html
[43] Ted. C. Fishman, *China Inc.*, Scribner Press, 2005
[44] Ibid.
[45] Thomas Lum and Dick K. Nanto, CRS Report for Congress: China's Trade with the United States and the World, January 4, 2007
[46] Thomas L. Friedman, *The World is Flat: A Brief History of the Twenty-First Century*, Farrar Straus & Giroux, 2005
[47] Douglas A. Irwin, *Free Trade Under Fire*, Princeton University Press, 2002
[48] United States International Trade Commission, "Harmonized Tariff Schedule of the United States (2010) (Revision 1)"

Chapter 6 – Energy Independence and Environmental Sustainability

[49] Paul Roberts, *The End of Oil: On the Edge of a Perilous New World*, Houghton Mifflin Company, 2004
[50] Matthew R. Simmons, *Twilight in the Desert: The Coming Saudi Oil Shock and the World Economy*, John Wiley & Sons, Inc., 2005
[51] Richard Mullins and Joaquin Sapien, "Wasting Away," The Center for Public Integrity, 2007
http://projects.publicintegrity.org/superfund/report.aspx?aid=853
[52] National Academy of Sciences, "Hidden Costs of Energy: Unpriced Consequences of Energy Production and Use," 2009 The National Academies Press
[53] Lisa Wangsness, "New Offshore Drilling Not A Quick Fix, Analysts Say," *The Boston Globe*, June 20, 2008
[54] Vladimir Socor, "The Shtokman Gas Deal: An Initial Assessment of Its Implications," *Eurasia Daily Monitor*, Volume: 4 Issue: 138, July 17, 2007
[55] For a working draft of the Alaska's Clear and Equitable Share bill, go to: **http://www.gov.state.ak.us/pdf/773-08-0014%20DRAFT%20oil%20tax%20bill%2010-1-07.pdf**
[56] Jon Hilsenrath, "Cap-and-Trade's Unlikely Critics: Its Creators", *Wall Street Journal*, August 13, 2009
[57] Ibid.

[58] Lejla Alic, "Trends in US Residential Natural Gas Consumption," Office of Oil and Gas - Energy Information Administration, June 2010

[59] Energy Information Administration (EIA), "Household Vehicles Energy Use: Latest Data & Trends – Based on Augmentation of the January 2004 Release of the 2001 National Household Travel Survey conducted by the US Department of Transportation and Other Relevant EIA Data," EIA – US Department of Energy, November 2005

[60] The Pew Charitable Trusts "Who's Winning the Clean Energy Race? Growth, Competition and Opportunity in the World's Largest Economies," 2010

Chapter 7 – The Federal Budget

[61] Congressional Budget Office, "The Budget and Economic Outlook: Fiscal Years 2011 to 2021," CBO January 2011

[62] Ibid.

[63] Ibid.

[64] In the FY2008 budget, over 62 percent of earmark spending went into the defense budget. In FY2009 the figure was over 56 percent and in FY2010 the figure was 42 percent. Prior to FY2008 it is difficult to determine an exact percentage because earmark spending did not have to be publicly disclosed. These figures are available from Taxpayers for Common Sense (www.taxpayer.net).

[65] Marc-André Gagnon and Joel Lexchin, "The Cost of Pushing Pills: A New Estimate of Pharmaceutical Promotion Expenditures in the United States," PLoS (Public Library of Science) *Medicine*, January 3, 2008, http://www.plosmedicine.org/article/info:doi/10.1371/journal.pmed.0050001

[66] Congressional Budget Office, "The Budget and Economic Outlook: Fiscal Years 2011 to 2021," CBO January 2011

[67] *Report of the Task Force on a Unified Security Budget for the United States, FY 2010* (Washington DC: Institute for Policy Studies and Foreign Policy in Focus, September 2009)

[68] *Debt, Deficits, and Defense – A Way Forward*; Report of the Sustainable Defense Task Force; June 11, 2010

[69] Ibid.

[70] Center for American Progress, "A Thousand Cuts – What Reducing the Federal Budget Deficit Through Large Spending Cuts Could Really Look Like," September 2010
[71] James Wood Forsyth Jr., Col. B. Chance Saltzman, Gary Schaub Jr., "Remembrance of Things Past: The Enduring Value of Nuclear Weapons," *Strategic Studies Quarterly* (Spring 2010)
[72] Department of Energy Fiscal Year FY2011 Budget Justification Book Volume 1 (Washington DC: Department of Energy, February 2010)
[73] Christopher E. Paine, *Weaponeers of Waste: A Critical Look at the Bush Administration Energy Department's Nuclear Weapons Complex and the First Decade of Science-Based Stockpile Stewardship*, Natural Resources Defense Council, April 2004
[74] Congressional Budget Office, *Budget Options, Volume 2* (Washington DC: Congressional Budget Office, August 2009)
[75] *Debt, Deficits, and Defense – A Way Forward*; Report of the Sustainable Defense Task Force; June 11, 2010
[76] Troop reductions and savings mentioned here are my own suggestions and estimates, drawn from various suggestions made by the Sustainable Defense Task Force and economists from the CATO Institute all included in: *Debt, Deficits, and Defense – A Way Forward*; Report of the Sustainable Defense Task Force; June 11, 2010
[77] *Debt, Deficits, and Defense – A Way Forward*; Report of the Sustainable Defense Task Force; June 11, 2010
[78] Robert O. Work, *The U.S. Navy: Charting a Course for Tomorrow's Fleet* (Washington, DC: Center for Strategic and Budgetary Assesments, 2008)
[79] *Debt, Deficits, and Defense – A Way Forward*; Report of the Sustainable Defense Task Force; June 11, 2010
[80] This litany of problems is detailed in a GAO report: Michael J. Sullivan, *V-22 Osprey Aircraft: Assessments Needed to Address Operational and Cost Concerns to Define Future Investments* (Washington DC: Government Accountability Office, 23 June 2009)
[81] *Debt, Deficits, and Defense – A Way Forward*; Report of the Sustainable Defense Task Force; June 11, 2010
[82] Congressional Budget Office, *Budget Options, Volume 2* (Washington DC: Congressional Budget Office, August 2009)

[83] *Debt, Deficits, and Defense – A Way Forward*; Report of the Sustainable Defense Task Force; June 11, 2010
[84] Ibid.
[85] Congressional Budget Office, *Budget Options, Volume 2* (Washington DC: Congressional Budget Office, August 2009)
[86] *Debt, Deficits, and Defense – A Way Forward*; Report of the Sustainable Defense Task Force; June 11, 2010
[87] Ibid.
[88] *Report of The Tenth Quadrennial Review of Military Compensation, Volumes I & II* (Washington DC: Department of Defense, Undersecretary of Defense for Personnel and Readiness, February 2008, July 2008)
[89] Congressional Budget Office, *Budget Options, Volume 2* (Washington DC: Congressional Budget Office, August 2009)
[90] *Debt, Deficits, and Defense – A Way Forward*; Report of the Sustainable Defense Task Force; June 11, 2010
[91] Congressional Budget Office, "The Budget and Economic Outlook: Fiscal Years 2011 to 2021," CBO January 2011
[92] Congressional Budget Office, "Social Security Policy Options," CBO July 2010
[93] Social Security Administration - http://www.ssa.gov/history/lifeexpect.html
[94] Congressional Budget Office, "Social Security Policy Options," CBO July 2010
[95] Ibid.

Chapter 8 - Conclusion
[96] This figure is for tax year 2007 and was computed using IRS figures available at: **http://www.irs.gov/pub/irs-soi/07intop400.pdf**